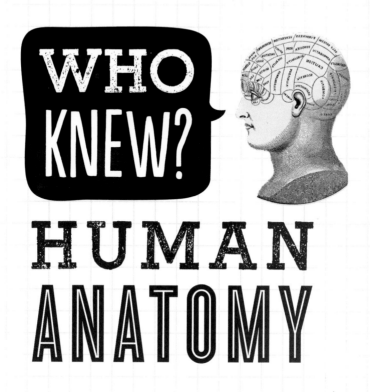

WHO KNEW?

HUMAN ANATOMY

Sophie Collins

PORTABLE
PRESS

Portable Press
An imprint of Printers Row Publishing Group
A division of Readerlink Distribution Services, LLC
10350 Barnes Canyon Road, Suite 100, San Diego, CA 92121
www.portablepress.com

Correspondence concerning the content of this book or permissions should be addressed to Portable Press, Editorial Department, at the above address. Author and illustration inquiries should be addressed to The Bright Press, part of The Quarto Group, Level 1, Ovest House, 58 West Street, Brighton, UK, BN1 2RA.

Publisher: Peter Norton
Associate Publisher: Ana Parker
Editor: Traci Douglas
Publishing Team: Vicki Jaeger, Lauren Taniguchi, Stephanie Romero

Conceived and designed by The Bright Press, part of The Quarto Group, Level 1, Ovest House, 58 West Street, Brighton, UK, BN1 2RA

Publisher: Mark Searle
Creative Director: James Evans
Managing Editor: Jacqui Sayers
In-house Editor: Judith Chamberlain
Project Editor: Lucy York
Cover and Interior Design: Clare Barber
Text: Sophie Collins

ISBN: 978-1-68412-786-3

Library of Congress Cataloging-in-Publication Data available on request.

Printed in China

23 22 21 20 19   1 2 3 4 5

"The first and simplest emotion which we discover in the human mind, is curiosity."

Edmund Burke

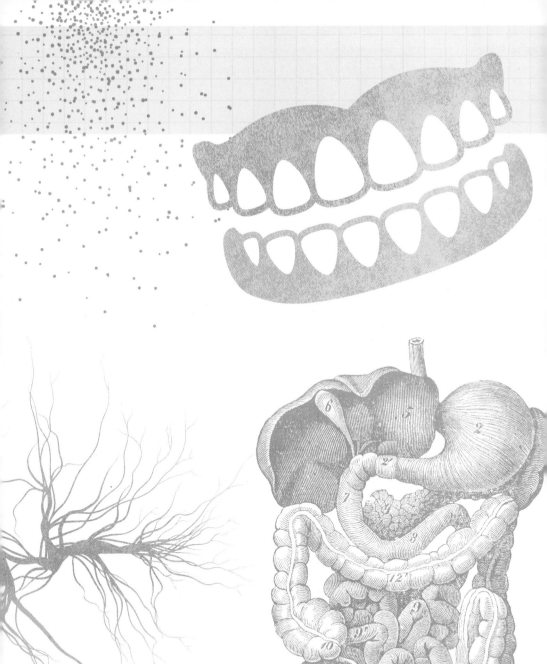

CONTENTS

INTRODUCTION

You probably live in one—if you're reading this, in fact, you certainly do—but how much do you actually know about it? Sadly, the human body doesn't come with an owner's manual, and even for the experienced owner it's full of surprises. This is the book that steps back, looks at a big list of the things you really want to know, and both asks and answers all those peculiar questions that come to you in odd moments, from whether or not it would be possible to fingerprint a newborn baby to whether fear could really turn your hair white.

Ten chapters take you on a virtual tour, from conception (Birth and Before) to last rites (Death and After). Between the two, you'll look, among other things, at Fashionable Bodies (Could your body be 100 percent tattooed?), Internal Affairs (Is it true that the microbiome in your gut could actually cause depression?) and Unexpected Events (Could a human being spontaneously combust?). Along the way, you'll pick up all kinds of information. Some of it may be of practical use (Could sex cure a headache?), some of it more theoretical (What's the weight of a human soul?), but every bit of it will be entertaining and often unexpected, too—not to mention boosting your personal standing by no small amount at the next quiz night you attend.

Perhaps the most surprising thing you'll discover is how many different habitats one human body can contain. The analogies come thick and fast depending on which bit of it you're scrutinizing. There's the crowded city of

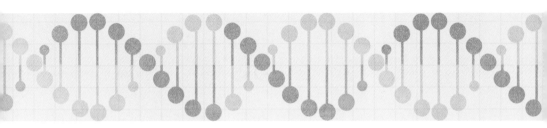

the gut, where millions of bacteria dwell in an uneasy truce in a humid, active environment; and there's the Wild West of the circulatory system, where strangers, whether bacteria or viruses, will be literally engulfed and dismantled by crack teams of "sheriff" cells. Then there's the electric field of the brain, which may run on a puny 12.6 watts, but within which trillions of synapses are constantly firing their messages to and fro, ensuring that you can walk and talk (and even think) at the same time.

There's something particularly appealing about the idea that, as you read this, your body—that extraordinarily rich and complicated machine—is going about its business, in the main without fuss, dealing with the majority of problems before you, its owner, are even conscious of them. It's a jungle in there. Now read on.

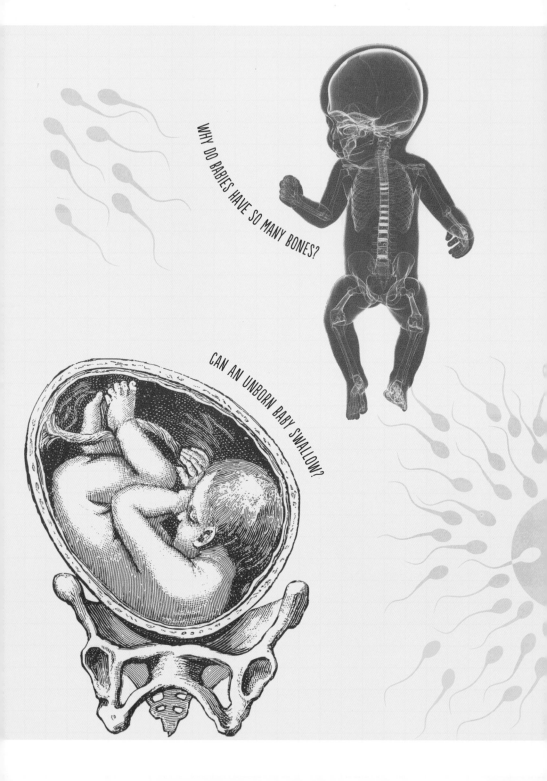

WHY DO BABIES HAVE SO MANY BONES?

CAN AN UNBORN BABY SWALLOW?

WHY DO HUMANS HAVE THE LONGEST LABOR?

WHERE'S THE SAFEST PLACE IN THE WORLD TO GIVE BIRTH?

BIRTH AND BEFORE

HOW BIG IS A HUMAN EGG?

Why are unborn babies sized by sorts of fruit?

Most adults are weight-literate, so why does almost every guide to the developing fetus, from the moment the first cell divides into two in the womb to the birth of a full-term baby, offer cutesy weekly updates of your baby's likely size in terms of fruit and vegetables?

From Sesame Seed to Watermelon

The medical sizing of unborn babies, due to their curled-up position in the womb, is given by a measurement taken from the crown of the head to the lower curve of the bottom, called the CRL (or crown–rump length). Development checks taken in the course of a pregnancy compare CRL with GA (gestational age) to confirm that the baby is growing at the rate expected.

Move into the popular arena, though, and baby books and magazines seem to have found this stark means of measurement too unappealing to pass on. It's not known when the first comparison with fruit was made and a mother-to-be was told that her 15-week fetus was about the size of an apple, but by the mid-1980s most baby books had caught on, and today almost every source of information for pregnant women, from magazines to posters to websites, will have a guide to show them which milestone grocery item their baby matches in size this

DOES FRUIT FUEL SMARTS?

Maybe it's not so much the "fruit" coming out as the fruit going in that counts. An extensive study made in 2016 by the University of Alberta concluded that increased fruit intake in the diets of expectant mothers seemed to have a positive effect on the brainpower of their babies. Around six portions of fruit a day, set against none or very few, resulted in children who subsequently scored an additional six or seven points in IQ tests.

week. Some are more reassuring than others; while a grape or an orange sound comfortable enough, a pineapple raises rather spiky associations, and a coconut sounds horribly . . . unyielding. Could most people confidently envisage, let alone size, a kumquat or a rutabaga? And by the time a pregnant woman hits the 40-week mark, she may be less thrilled to hear that her baby is now the size of a "medium" pumpkin or watermelon.

Comfortably Curvy

Nonetheless, it seems that the fruit vs. fetus sizing chart is here to stay. Some medics have offered up the explanation that pieces of fruit have mostly wholesome, desirable associations: healthy, organic, and (in the main) curvy in shape. A number of alternative comparison charts have been created, some more ironically than others (comparing fetus size at various stages to a smartphone, say, or a burger), but none has caught on to the same extent, even if the idea of birthing a watermelon might be daunting to the most enthusiastic mother-to-be.

Why do babies have so many bones?

They may be small, but babies notch up a total of 300 bones apiece—considerably more than adults do. By the time they grow up, the total will have reduced: grown-up humans have a comparatively modest 206 bones. What happens to the extras?

Bone vs. Cartilage

The effort of the human birth process has been described as being like trying to squeeze a grapefruit out of your nostril—and humans have the most protracted births of any mammal. In order for it to be actually possible for the mother to push the baby through the birth canal, the newborn needs to be very bendy indeed. So, instead of rigid bone, a large number of those 300 baby "bones" are made up of cartilage, which is a much softer and more elastic substance. As an adult, pads of cartilage will remain at the ends of your long bones, cushioning the joints, as well as in your ribs, ears, and nose, but most of it will have fused into fewer, larger pieces of bone.

Hardening-Up

The technical name for this hardening-up is endochondral ossification. Cells called osteoblasts build bone, gradually replacing the cartilage from the inside out. The baby cartilage "skeleton" offers the template on which bone will grow, but the cartilage doesn't itself become bone.

The key element the osteoblasts need for the process is plenty of calcium and vitamin D, which aids the absorption of calcium. This is why so much emphasis is placed on calcium intake for pregnant women and nursing mothers. After birth, calcium is passed on to the infant in breast milk, and babies and children, unlike adults, can "bank" calcium stores.

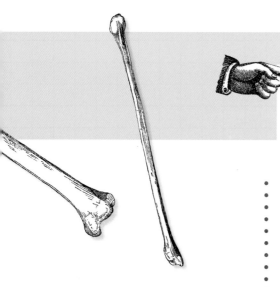

BONE-HEADED

Most people have felt the softness of baby bones for themselves when they've touched the area around the anterior fontanelle, or soft spot, at the top of a baby's head. This is the most obvious sign that the five overlapping plates that form the baby's skull haven't fused yet. There's also a second soft spot, the posterior fontanelle, at the back of the skull. Although the fontanelles are both closed by the age of three, the skull's bony plates are held together by flexible structures called sutures, and won't finally fuse into a single structure until the brain has reached its full size, when the child is around 7 years old.

How Long Does it Take?

Babies stay bendy for quite a while. If you've ever marveled at the resilience of a child at the crawling-to-walking stage, and its ability to fall constantly without hurting itself, the reason is its retained flexibility. The solid-bone skeleton doesn't form all at once; the growing of bones is a long process that isn't completed until a person hits their early twenties. And bones change even in adulthood—bone can grow to support and mend injuries. Your skeleton also "maintains" itself, replacing upward of 5 percent of its bulk with fresh bone every year.

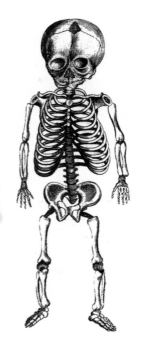

Could you fingerprint a newborn?

The only crime that most new babies could be accused of is parental sleep-theft, but if you needed to, would it be possible to fingerprint a baby, and would they have unique fingerprints, as adult humans do?

One-Offs

The short answer is yes. A baby's fingerprints develop in the womb, and by the time the fetus is 6 months old, they're completely formed. First steps toward the finished prints begin when the fetus is between 2 and 3 months old, and its fingers are developing. By the end of the third month, there are pads on its fingertips—somewhere for fingerprints to grow. At the 2-month point, the fetus has only two layers of skin, the inner, basal, and the outer, periderm, layer. Between 2 and 3 months, however, a third layer of skin begins to form, and it will be in this layer that hair follicles eventually develop. The three layers, though, grow at different rates, causing wrinkles and folds and, ultimately, it's the disparity between the layers and the pressure as the baby touches the surfaces around it in the womb that will create its fingerprints. And because no baby's womb experience is identical to another's, the fingerprints will reflect the one-off journey. Compared with those on the tip of an adult finger, the patterns recognized by experts as comprising a fingerprint—the whorls, arches, and loops—will be very shallow and faint on a fetus, but they are already present in the form in which they'll persist into adulthood.

Ten Tiny Fingers, Ten Tiny Toes

If you're imagining a baby's fingers being pressed first onto an ink pad and then, one by one, onto a piece of paper, in the way you've seen fingerprints being taken in police procedurals on

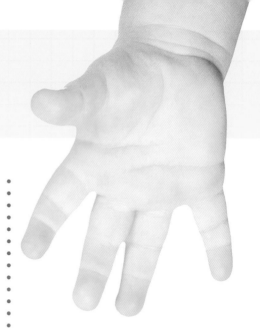

television or in movies, you can dismiss the idea. Young babies' fingerprints are far too faint and subtle to be captured by the old-style ink-and-paper process. You'd need a specialized high-definition scanner to capture them, and possibly some extra enhancement for a reliably clear image. (Incidentally, a baby's toe prints are just as unique, although arguably even harder to scan.)

And why might you want to take a small baby's fingerprints? Experiments have been done in the United States to see how easily a biometric database of children's fingerprints might be put together. Having a complete record would mean that any hospital confusions of the who-got-the-right-baby kind (rare, but not unknown) could be easily resolved, that missing or stolen children could be reliably traced, and that the need for many DNA tests (comparatively slower and more expensive) would be eliminated.

Is it Reliable?

At the moment, the use of a sensitive scanner assisted by an algorithm to refine the results isn't 100 percent reliable; however, in a study of a large number of children made in Agra,

India, while children of 6 months and over proved to be easy to fingerprint and trace (with results offering over 99 percent accuracy), prints taken from children under 1 month old fell to just over 50 percent accuracy in identification. At the moment, the jury is out on whether scanning the iris of the eye might ultimately prove both easier to do and more reliable.

Where's the safest place in the world to give birth?

A UNICEF study published in 2018 looked at births in 195 countries in 2016, and established that Japan is the safest country of all in which to give birth, closely followed by Iceland and Singapore.

Top Ten, Bottom Ten

While the world has grown safer for children under the age of five over the last 20 years, the same isn't true for newborns—or for their mothers.

While in Japan, there's a less-than-one-in-a-thousand chance that a baby will die in its first month of life, at the other end of the league table, in Pakistan there's one death for every twenty-two live births. The countries with the worst UNICEF statistics tend to be poor, without much in the way of resources for either education or medical care, and are also often war-torn and politically unstable. Afghanistan, Somalia, and the Central African Republic featured just above Pakistan in the bottom ten. However, of the global high-income countries, neither the United Kingdom nor the United States had much to boast about. According to UNICEF, the United Kingdom came thirtieth in the league table, and in the United States one baby in every 3,800 dies during birth. Why? The best-for-birth countries tended to have strong social structures around birth, subsidized healthcare services, and a high number of health professionals per head of the population. Overall expenditure

BADGE OF HONOR

Not only is Japan the safest country in which to birth a baby, it also offers some impressive side benefits for both pregnant women and new mothers. Mothers-to-be are given a special badge at their first check-up, requiring others to prioritize them in queues and to give up subway seats to them—and, after the birth, new parents get a one-time payment that can be as high as 420,000 yen (that's just under $4,000, or nearly £3,000). On the slightly less enviable side, pain relief isn't widely used at a "normal" Japanese birth—many hospitals don't even offer the option of an epidural. And pregnancy weight gain is monitored very closely, with a total gain of no more than 22 pounds being considered acceptable. Having given birth, most mothers will stay in the hospital with their babies for at least 5 days of aftercare, longer if they've had a caesarean.

of GDP on childbirth services seemed to affect the tables less than you might expect, though: The United States spends 16.6 percent, against Japan, at a little under 11 percent, while the United Kingdom spends under 10 percent. In some cases, levels of highly technical medical intervention appeared to influence the statistics to negative rather than positive effect.

Streetwise

When it comes to unsolicited advice from strangers on the street, every country has its cultural norms. In Japan, expectant mothers are most likely to be told to keep their bellies and ankles warm, in case the baby catches cold, while in Southern India women are encouraged to get below their pre-birth weight as soon as possible after birth, to drain them of the toxic fluids of pregnancy—recent mothers who are still on the rotund side may well be scolded by older ladies.

Can an unborn baby swallow?

Even while it's still in the womb, a baby has quite a lot of ways both of occupying itself and making its presence felt. It can kick, somersault, hiccup, cry, and suck its thumb. What other abilities does it have? Can it swallow?

Not for the Squeamish

For the first 11 weeks of life, the fetus's embryonic "mouth" (although not really recognizable as such at this stage of development) has a layer of cells, called the buccopharyngeal membrane, over it. This seals the area, meaning that it's nil by mouth for the growing baby. At around 12 weeks, though, the membrane gradually ruptures, kickstarting the swallowing reflex in the baby.

It's already able to urinate—kidney function gets going at around 8 weeks and the kidneys are well developed by the sixteenth week. And since their activities are limited to the amniotic sac and the fluid inside it, once a baby can both swallow and pee, most of what it swallows will end up being its own urine.

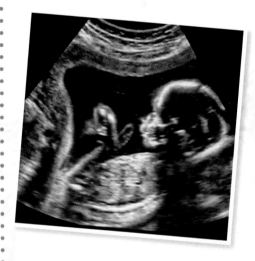

Swimming in . . . what?

At the beginning of a pregnancy, amniotic fluid is made up of water and a small amount of salts. Over time, though, as the baby develops further and becomes better at swallowing and increasingly efficient at urinating, more and more of the amniotic fluid turns to urine.

At regular points during the pregnancy, doctors check that the level of amniotic fluid in the sac doesn't drop. Once they know that the fetus is swallowing, it's essential to know that it's also eliminating—it offers important proof that the kidneys, which must be fully functional by the time the baby is born, are working efficiently. And while the liquid that the growing fetus ends up swimming around in is technically urine, it's very different from adults', or even children's urine. It is much less concentrated and free of urea (which remains in the placenta, and which is the substance that gives regular urine its sour, ammonia smell).

NUMBER TWOS

Be thankful for small mercies, though: Babies don't in general move their bowels until after birth. The first motion consists of meconium—a thick, treacly substance that has accumulated in the baby's gut and consists of things that couldn't be eliminated by urinating: dead cells, hair, sticky mucus, and so on. If a baby does eliminate solids in the womb, as happens in around 13 percent of pregnancies, there's an immediate danger to its well-being, as meconium, if swallowed, may choke the baby or cause an intestinal block. Usually, though, the meconium is saved as a messy, sticky (although at least not stinky; it has almost no smell) surprise for the diaper-changer shortly after the baby is born.

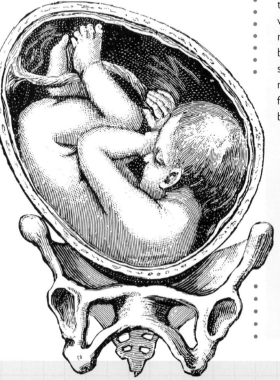

Why do humans have the longest labor?

It is believed, by humans at least, that *Homo sapiens* has it toughest when it comes to labor. The huge head of the infant and the narrow hips of the mother are cited as the reason that labor is so hard.

If Only They Could Talk

It's true that walking upright has resulted in a tricky, angled birth canal, whereas most of our ape relatives give birth through something more closely resembling a short, straight pipe. And human labor is generally longer than that of most other mammals—averaging between 10 and 20 hours, in contrast to the around 2-hour labor of most other primates—because the size of the head means that the cervix needs to be very widely dilated before the mother can push the baby out.

Nonetheless, while no one would dispute that there's a reason it's called "labor," a brief study of the birthing procedures of some other mammals doesn't necessarily win the toughest-labor argument for humans. Most mammals show signs of discomfort as they give birth, and many display symptoms of pain. True, some do seem to have it relatively easy—a sow might have a litter of a dozen or more piglets, but in many cases they will each weigh a tiny proportion, perhaps just a 500th, of her own weight, so it's fair to argue that they do "just slip out." Whales and other *cetaceans* benefit from natural water births: The water supports the weight of the mother, and this helps her with the hard work of labor.

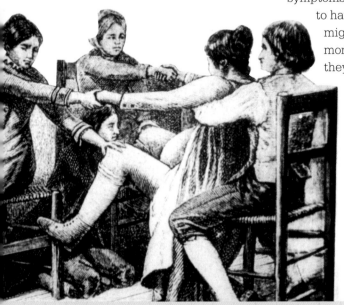

CHALLENGING THE OBSTETRIC DILEMMA

For many years it was believed that a human baby grows until just before the point at which its head would be too big to fit through the pelvis, and that the pelvis is relatively narrow to allow the mother to move efficiently on two legs, so it couldn't evolve to be wider. Known as the Obstetric Dilemma, the argument was widely accepted until, in 2012, a study came along to open the door to new theories. Holly Dunsworth, an anthropologist at the University of Rhode Island, argued that women could easily have developed wider hips without compromising their efficiency of movement. Instead, she hypothesized, humans give birth at the point at which the mother is no longer able to meet her baby's high nutritional needs in the womb: It's simply easier to cater for the baby outside her own body.

Worst-Case Scenario

One mammal, though, would surely win any cross-species birth horror story contest. The poster child for poor design when it comes to a vaginal delivery is the spotted hyena, whose peculiarity is the extraordinarily high levels of male hormones that all cubs, male or female, are exposed to in the womb. One result of this is that both sexes have what appears to be a penis; in the case of the female, it is actually an elongated, tube-shaped clitoris down the narrow length of which she must give birth. It's a tortuous process, and first-time mothers often have great difficulty in birthing their cubs—cubs themselves so turbocharged with testosterone that they often start fighting each other moments after birth.

How big is a human egg?

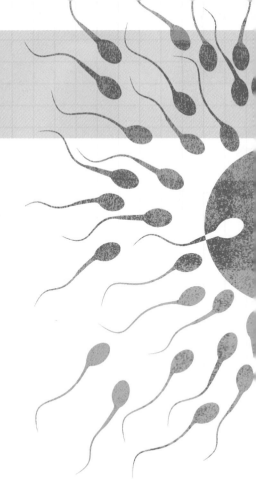

Measured against the other cells in your body, a human egg is huge. Of course, that's in cell terms; everything is subjective, and at about a tenth of a millimeter in diameter, which is just about visible to the human eye if you were peering very hard, it may not look very big. But in cell terms, it's a monster, at least sixteen times as big as a human sperm, for example, and four times the size of the average blood cell.

Packing a Punch

Why does the egg need to be so large? After all, every tiny sperm contains just as many chromosomes as are found in the egg's nucleus—so the chromosomes don't take up much space. And unlike, say, hen's eggs in which a lot of the available room is taken up by the yolk to feed the developing chick, human eggs don't have to be full of food. They just need a little to tide a fertilized egg over for a few days until it implants in the uterus lining—after which, the mother will supply nutrition on tap. So what else is in there?

Getting to Work

Once the egg has been fertilized by a sperm, it becomes a zygote—still a single cell, but one in which two entities are linked—the first cell of what will ultimately become a new person. While there's still more research to be done on the exact content breakdown of the egg, scientists know quite a lot, and the knowledge hinges on all the jobs that become urgent once the egg has become a zygote.

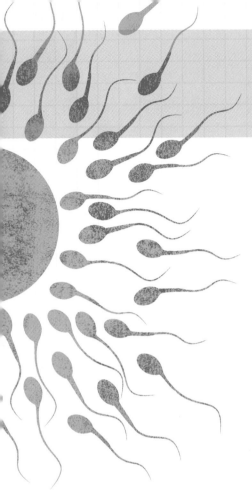

initial stages of cell division and sends messages to the cells, directing them as to what their future roles in the new embryo will be.

They know that there's a lot of ribonucleic acid (RNA) in there, and plenty of mitochondria (usually called the "powerhouses" of the cell, due to their ability to transform raw material into chemical energy). The latter help out by using existing oxygen and nutrients to make the energy that the fertilized egg will need, while the RNA acts as a kind of director to the egg's initial processes. It works on fusing the nuclei of egg and sperm together, and, after fertilization, it helps with the

A LIFETIME'S PRODUCTION

You've probably heard that a female baby is born with all the eggs she will ever produce already inside her. Newsflash: scientists are reassessing this in the light of the recent discovery of a previously unknown kind of stem cell in women's ovaries—which makes it at least possible that women are actually creating fresh eggs all through their fertile years. This may go a small way toward leveling the playing field, given that the average ejaculation can contain more than a billion sperm and that a young, fertile man can produce up to 300 million sperm cells every single day.

Why does pregnancy make a woman's feet bigger?

It may not seem all that surprising. After all, quite a few parts of the body grow bigger in the course of a pregnancy. But what makes the difference between temporarily swollen feet, and feet that seem to have grown permanently larger?

Relaxing Relaxin

Traditionally, the growth is blamed on the hormone relaxin. Relaxin loosens up the muscles and ligaments and softens the cervix, increasing your pelvic flexibility in preparation for giving birth. Increased levels of relaxin may also affect non-pelvic joints. With twenty-six bones, thirty-three joints, and plenty of ligaments in each one, human feet are prime candidates for increased stretching. But the natural water retention of pregnancy can also result in swollen feet and ankles, so it may be hard to tell whether or not the side effects of relaxin are to blame.

Other factors can include a higher-than-average weight gain—which is likely to broaden a person's feet as they cope with balancing the increased load—and a lot of standing around. A mother will only really know whether larger feet are there to stay a few months after birth, when they're still measuring up at their new—bigger—size.

How Much Bigger?

When feet do grow larger, it's usually at most by one full size, although the arch of the foot may also flatten out a little. Studies haven't been conclusive, but the Bigfoot effect appears to occur in around one in four women, and only in their first pregnancy—which must be a relief to anyone who is planning a large family and whose feet are already one size larger just one birth into the process.

Does your birth month affect your performance in school?

Monday's child is fair of face, goes the rhyme, but is it true that the month—rather than the day—of your birth may have a real bearing on how well you do at school? And, if so, why?

Summer Babies, Winter Babies

The school year starts in fall, so a child with an August birthday may just scrape into the fall class, although they will be the youngest in the year, while a September birthday means a wait until the next year, and the advantage of being oldest in the class, so theoretically more ready for everything that education has to throw at them.

Another argument holds that fall or early winter babies thrive both physically and mentally because their mothers were pregnant during spring and summer, enjoying better weather and more sunlight, and, in theory, at least, fresher, more nutritious food.

A Head Start

Are either of these supported by research? Well, yes and no. It does seem to be true that children who start school a little older get a head start on education, possibly because their brains are (slightly) more developed than those of their younger classmates and they find it easier to concentrate, an ability that tends to increase with age. One British study, undertaken in 2013 by the Institute for Fiscal Studies, found that August-born children were 2 percent less likely to go to university than their September-born contemporaries—a difference, although not a huge one. And in terms of sporting success, another United Kingdom study found that children born in September, October, and November are more than three times as likely to be picked by professional sports academies than those born in June, July, or August.

When was the first Cesarean section?

You may know the Greek legend in which the god Apollo cuts his son, Asclepius, from the dead body of Coronis, as she lies on her funeral pyre. This gory story is the first recording of a cesarean section. Asclepius would go on to become the god of medicine.

Story and Substance

The *Suda*, an encyclopedia of the ancient world produced around the second half of the tenth century AD, attributed the name to the birth of Julius Caesar, saying that "When his mother died in the ninth month, they cut her open." It goes on to say that *caesar* is the Roman word for dissection—probably a reference to the Latin verb *caedere*, "to cut." There's a muddle about which Caesar is referred to in the story—Julius Caesar's mother was alive well into his adulthood. The operation may have been named for another Caesar—it wasn't an uncommon name in Roman times.

From Pigs to People

The first written account of an operation survived by both mother and child comes from Switzerland in 1500. Jacob Nufer cut his wife's abdomen open in order to extract the baby (and then sewed her up again) after the intervention of no fewer than thirteen midwives had failed to help her to give birth. His unglamorous profession—gelding pigs—may have given him some rudimentary anatomical knowledge. Whatever the reason, it worked. The lady reportedly went on to have five more children.

OINK

BIRTH AND BEFORE

From the moment they're born, humans acquire knowledge at an astonishing rate. Now that you're all grown up, prove how much you learned in this chapter by taking this quiz.

Questions

1. If you're measuring the size of an unborn baby, what does CRL stand for?

2. New babies have more bones than adults. How many more?

3. Are a new baby's fingerprints unique?

4. A human egg is around twice the size of a human sperm—true or false?

5. Why might a spotted hyena dread giving birth?

6. Relaxin is the hormone that helps you drift off to sleep—true or false?

7. The first Cesarean section is said to have been performed by a Greek god. Which one?

8. Japan gives every expectant mother a special badge. What's it for?

9. What is meconium?

10. Birthday in August or September—which gives you an advantage in school?

Turn to page 212 for the answers.

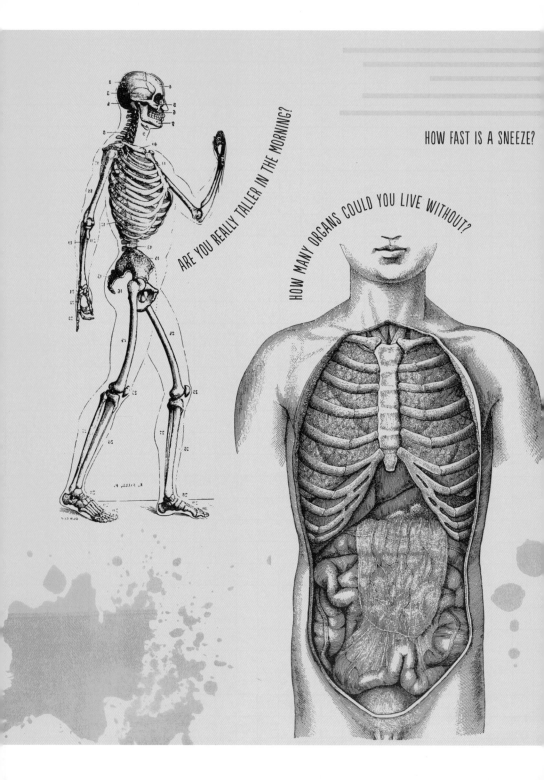

ARE YOU REALLY TALLER IN THE MORNING?

HOW FAST IS A SNEEZE?

HOW MANY ORGANS COULD YOU LIVE WITHOUT?

FASTEST, LONGEST, GREATEST, STRONGEST

HOW MUCH BLOOD, SWEAT, AND TEARS WILL YOUR BODY PRODUCE IN A LIFETIME?

HOW MANY MILES OF TUBING DOES THE AVERAGE BODY CONTAIN?

How fast is a sneeze?

The old rhyme goes: "Coughs and sneezes spread diseases," and there's a reason for this. The strong propulsion of a sneeze makes it an effective spray of your microbes out into the world. But how powerful is the everyday "achoo"?

Faster Than a Speeding Bullet?

Not quite. A bullet flies at around 1,700 mph. But a sneeze can still pack quite a punch for something that started as an irritation of the mucus membranes of the nose and turned into a rapid expulsion of air through the mouth. The fastest sneeze ever recorded under controlled scientific conditions traveled at 102 mph: Way slower than a bullet, but by far the quickest of our bodily expulsions (even the fastest cough racks up only around 60 mph). When the U.S. television show *MythBusters* conducted some on-air tests in June 2010, they concluded that the average sneeze comes out at around 37 mph—still quite speedy.

Bless You!

When you find out the true facts about the pathogens spread by a sneeze, you'll be less inclined to bless the offender. A study carried out in 2014 at the Massachusetts Institute of Technology's department of Civil and

Environmental Engineering uncovered some alarming statistics. Not only does a sneeze—given the less-than-catchy title of "multiphase turbulent buoyant bubble"—propel largish droplets of liquid into the sneezer's surroundings, but thousands of much smaller damp gas particles—invisible to the naked eye—are released, too.

How Far?

The buoyant bubble, incorporating up to 100,000 germs, can float as far as just over 26 feet. Worse, the pathogens it contains can remain in the air for as long as 10 minutes. And perhaps worst of all, the tiny particles of damp gas tend to move upward toward the ceiling—just where most buildings (schools, offices, hospitals) site the ducts for their ventilation systems. Which means that the effects of an energetic sneeze could impact people much further away than just those in the sneezer's immediate vicinity.

Sneeze Etiquette

So if you feel a sneeze coming on, it's more important than you may have thought to cover your mouth. And not with your hand, which, unless you carry a pack of antiseptic wipes

with you for a post-sneeze hand-wipe, may afterward spread even more pathogens about when you touch the surfaces around you. If you don't have a tissue, another piece of research suggests that the best thing to do is to use your inner elbow to "catch" the sneeze, as this is the part of your body that's least likely to come into contact with other surfaces.

How many organs could you live without?

Wе all know someone who's had their appendix out (and good riddance, they may feel: An inflamed appendix is famously painful, and it's an organ with no known function anyway), but how many other organs could you live without altogether?

Optional Organs

You can lose one kidney without too many problems—provided that the other one is fully functional, it will do its best to keep up with the work, although you can help it by following a healthy diet and eschewing alcohol. The same goes for the lungs. Typically, a person uses only around 70 percent of their potential lung capacity with two lungs, so losing one doesn't mean losing half their full capacity, although the remaining lung will have to work harder to compensate.

Of course you can also manage without a uterus (in a woman) or testicles (in a man), although you won't be able to have children. And it's possible, too, to cope without a complete digestive system—you could live without your stomach or much of your colon, though not without discomfort or inconvenience. Taking out your

spleen would mean an increased susceptibility to infection because it's a part of your immune system, and if your gallbladder had to be removed, you'd no longer have the extra bile that it stores to help you digest fat, so you'd have to avoid excessively fatty meals.

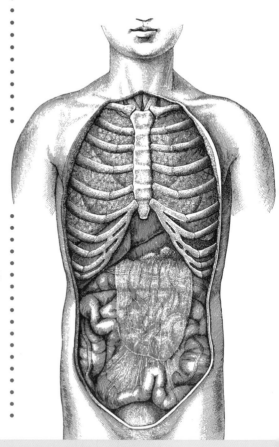

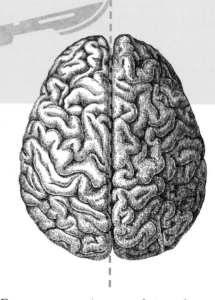

You would be exceptionally unlucky to lose all, or even most, of these organs—but theoretically, at least, you would be able to survive.

Half a Brain

You would probably be able to handle the suggestion that someone remove, say, your gallbladder with reasonable equanimity. But what if someone were to propose taking out a large chunk of your brain? Much less acceptable. Yet there's an operation called a hemispherectomy that removes one hemisphere of the brain, and patients can live quite well for years with just the remaining half. The first hemispherectomy was performed at Johns Hopkins University, Maryland, in 1928, and the operation has since become a specialty there. It's a last-resort operation, usually carried out on patients who suffer so badly from seizures that daily life has become impossible, and the huge majority of patients are young children. John Freeman, an eminent professor of neurology at Johns Hopkins, quipped "You can't take more than half. If you take the whole thing, you've got a problem." Even the operation to remove half is not without side effects, although they're not so bad as one might suppose: Patients usually lose motor function on the opposite side of the body from where the brain was removed, and may have speech problems, but can generally operate normally in most other ways. What happens to the skull space left by the vacated brain? It doesn't stay empty; the obliging body quickly fills it with fluid.

"You can't take more than half. If you take the whole thing, you've got a problem."

How much blood, sweat, and tears will your body produce in a lifetime?

In 2015, the World Health Organization set global average life expectancy at 71.4 years. What's the volume of blood, sweat, and tears that such an "average" person might produce within their lifetime?

Sweating Like a ... Human?

You may have read that humans sweat as much as 2 pints of liquid a night. This may be pleasant reading for mattress salespeople, but can it be true? Unless you suffered from night sweats (sleep hyperhidrosis), you'd be unlikely to sweat that much while you were resting. It can be a different story when you're active, though. A study at the University of Ottawa found that an active person who included 45 minutes of intense exercise in their day might sweat as much as 4.25 pints across a 24-hour period. Allow that the same person might spend 50 years in this active pattern, making 77,562.5 pints and 21.4 years (at either end of their life—childhood and older age) sweating half as much, say 2 pints a day—adding a further 15,222 pints—and the total across their lifetime *could* work out at 93,184.5 pints. How much does that represent in the time-honored how-many-bathtubs calculation? With average bath capacity at almost 80 gallons, that's over 190 bathfuls of sweat.

Cry Me a River

After the sweat statistics, your tear production may come as an anticlimax. Various studies have given a wide range of results, although all found that women cry more than men, and for longer each time (one popular study has women crying an average of three and a half times a month, forty-two times a year, with men coming in at less than half of that). How many tears we shed has only been calculated by totaling the lubricating tears our eyes produce all the time (as opposed to the

much rarer event of emotional crying). Estimated total? Not anything close to a river, but still nearly 1,584 gallons of tears per lifetime—a mere twenty baths full.

What About Blood?

Unless you give it away (around a pint per donation) or have an accident, your blood tends to stay inside your body—a closed circulation system. The average adult will have between 9.5 and 11.6 pints of blood circulating around their body at any one time (variations in height and weight affect the overall volume, but blood usually makes up between 8 and 10 percent of your body weight). Surprisingly, children have the same volume of blood as adults by the time they're around 6 years old, although, since they're smaller, it's a higher percentage of their overall body weight (a newborn, by contrast, has well under a pint of blood in its veins).

How much blood will you produce in a lifetime? Cells in the blood are constantly dying and being replaced, so it would be hard to calculate accurately. However, if you give a pint of blood, your body will have replaced the volume within around 2 days. The contents—white cells, platelets, and in particular red blood cells—take longer to generate, so it may be several weeks before the ingredient balance is the same as it was originally. That's why most banks won't allow you to give blood more frequently than every 12 to 16 weeks.

❝ Don't even get us started on the stats for urine or saliva. ❞

How much skin do you shed in a day?

Your skin is the largest organ of your body and since—unlike, say, your lungs or your liver—it's on show all the time, it's also probably the one that you're most aware of. It constantly replaces itself, too, at a surprisingly fast rate.

Millions of Cells

Most people replace their outer skin layer completely every 4 weeks. The lower layer of your epidermis produces new cells that work their way to the surface, which is covered with a thick layer of dead cells. These are gradually shed—rubbing off against other surfaces, or blowing off in a current of air, or simply drifting off, invisibly, to be quickly replaced by freshly minted new cells.

How many cells does this mean you lose daily? Experts argue the count, but it's a huge number: Anything from a million to several times that. Of course, they're tiny, but over a year, other estimates feasibly claim that you'll lose more than 2.2 pounds of skin. Although it's not an appealing idea to us humans, dead skin cells do have some fans. They're called dust mites.

Little Helpers

Dust mites are arachnids, like spiders, only very, very small ones. They have some plus points: They don't bite, and they don't carry disease, so unless you have a dust-mite allergy, you're unlikely to be aware of them. They can be found wherever dust accumulates, preferring a moist environment without much light, and although they like things warm, they're not keen on direct heat. Upholstery and carpets are natural shelters for them, but a dust mite's ideal home is the mattress on your bed.

Your skin cells make up a fair proportion of the dust you find around the house. For most people, it's a hygiene issue: A clean house is dust-free. But some research published by the American Chemical Society in 2011 revealed that the skin component of dust may actually offer a healthy side benefit. The oil in human skin, called squalene, has been proven to absorb ozone, a common atmospheric pollutant. So dusty surfaces may actually be better for you than ones that gleam.

Mites' Likes

Most mattresses have plenty of shelter possibilities and the minute flakes of skin that you'll shed as you toss and turn in your sleep offer dust mites an all-you-can-eat buffet. What's more, your sweat and humid breath make the bed a relatively moist environment, and this ensures that shed cells will begin to decompose quite quickly. Mites don't like freshly shed skin (the keratin it's made from is too tough and dry for them), but prefer to wait for the spores in the air to alight on the moistened cells and for molds to begin to develop on their surface. The mites' mouths work on a kind of pincer movement, so the flat, thin shape of the cells, which makes them easy to grab, is also appealing.

If you've been good enough to offer dust mites a comfortable home, they will reproduce—fast, and in high numbers. If you've already heard enough, remember that you can disrupt their lifestyle by vacuuming your mattress and turning it frequently.

Which bit of your body works the hardest?

A number of your body organs have a Stakhanovite work ethic: They literally never rest. But in a crowd of workaholics, which bit of your body might win the prize for hardest worker? The three top contenders are the liver, the brain, and the heart.

The Liver

Galen, the Greek physician and philosopher who lived in the second century AD, gave the liver the prize: It was, he said "the principal instrument of sanguification," on which all the other major organs relied.

Was he right? The liver is the second largest organ (after the skin), weighing in at approximately 3 pounds. It filters your blood and keeps it clean, removing toxins and either posting them off to the kidneys (to be turned into urine), or turning them into bile and sending them to the gallbladder. It also deals to the best of its ability with any alcohol or drugs that enter your system. It makes cholesterol, which your body needs to make hormones, and keeps the levels of protein, fat, and sugar in your blood within reasonable limits. The liver is unique in that it can, to a degree, repair itself. It will also cope with a degree of abuse. That said, however, if you seriously damage your liver, with alcohol, for example, or an overly fatty diet, the only real fix is a transplant.

The Brain

It's a little lighter than the liver, and, perhaps surprisingly, it's around 70 percent water, but that doesn't diminish the fact that if your brain isn't working, neither are you. It also calls on 20 percent of the body's energy (although it represents only about 3 percent of the body's weight), of which two-thirds is used to keep your neurons firing and communicating with each other, while the remaining third is used to refuel the cells and keep them in

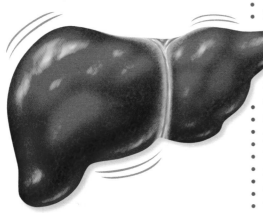

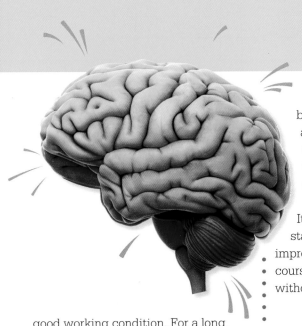

blood through your body with an average of around 100,000 beats a day, sending blood 12,000 miles through your circulatory system over a 24-hour period.

It would be hard to choose a stand-out winner from three such impressive organs. And the truth is, of course, that you couldn't function without any one of them.

good working condition. For a long time it was a popular misconception that we use only 10 percent of the brain's potential power at any one time, but this is now known to be false: Your brain is actually working on your behalf 24/7, whether you're active or resting. Over a 24-hour period, at one time or another, you will use 100 percent of its capacity.

The Heart

The size of two clenched fists and an indefatigable worker, a fit heart is one that's worked hard. And although it's a lightweight when it comes to heft—the average man's heart weighs a bit more than 10 ounces, the average woman's perhaps 2 ounces less—a healthy heart does an impressive job, pushing your

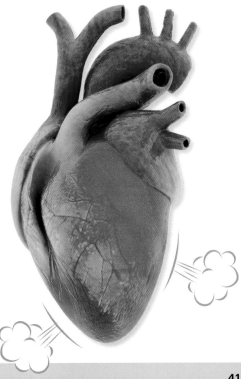

Which is the rarest blood type?

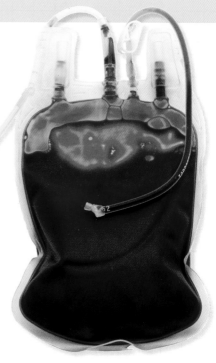

Usually the question "Which blood type are you?" is only asked in the context of someone needing a blood transfusion—but until comparatively recently, all blood was believed to have more or less the same qualities. It was only when blood was classified into different types that transfusion ceased to be an often-deadly lottery.

Putting Blood Back

The idea of "putting blood back" into a body that had bled heavily through injury or accident is an old one; it was first tried in the seventeenth century, shortly after the English physician William Harvey had discovered circulation. By the beginning of the nineteenth century, transfusions were regularly attempted, and sometimes they worked; at other times, though, patients had catastrophic reactions and the failure rate was still high.

The Discovery of Blood Types?

At the beginning of the twentieth century, Karl Landsteiner, a pathologist working at the University of Vienna, Austria, discovered the existence of blood types (he would later be awarded a Nobel Prize for his work). First he identified the A, B, and O groups, and a few years later his colleagues found the AB group. Blood cells can feature many different antigens, that is, substances that prompt the production of antibodies in the immune system. And at the end of the 1930s it was discovered that the presence or absence of a particularly powerful antigen, called the rH factor, could also affect your blood type—if you had the factor, you were positive, if not, you were negative. It's at this point that most people's knowledge of blood types ends: They know that O is the "universal" blood type, and that within each category there are "plus" or "minus" groups, but not really what that means.

It Takes All Types

Which group you belong to is crucial when it comes to either giving blood or having a transfusion. Someone given blood from the wrong group can have bad, sometimes very bad, problems. The blood may clot or coagulate, or the body may reject it altogether, leading to a catastrophic reaction. And even today, small, even tiny blood subgroups in which particular antigens are either present or absent are still being discovered. These are sometimes found in people who have a smaller gene pool as a result of being members of small societal groups.

- These small blood groups can create problems when it comes to a transfusion. "Bombay blood," for example, also known as the HH group, is so called because it was first written about in Mumbai in 1952 when doctors found two patients for whom no available blood in transfusion "worked." Eventually they were found to be unable to express the H antigen, which is present in all common blood types. This meant that although they could give blood to any of the other groups, they could only receive blood from a fellow member of the HH group.

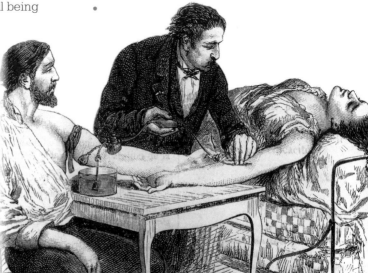

How many redundant parts are there in a human body?

Humans aren't the only species that have body parts that have become redundant: think of the ostrich's wings, for example. But we have our share of evolutionary hangovers from when our lives were lived very differently. You probably already know about your appendix and your coccyx—but they're not our only redundant bits.

Tall Tails

The tail bone, or coccyx, is just that—an extension at the base of the spinal column consisting of four or five fused vertebrae. Early in its development in the womb, the human fetus has a visible "tail," but it disappears around the eighth week of pregnancy. Very, very rarely a baby will be born with a short, vestigial tail, although it is usually made up of skin and fat, without a central bone.

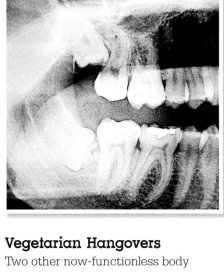

Vegetarian Hangovers

Two other now-functionless body parts are your wisdom teeth (around two-thirds of people have them) and your appendix (everyone is born with one). Both of these relate to a long-ago herbivorous diet. Early "salad" was made up of fibrous stems and greens, far tougher than the cultivated leaves we're familiar with today, and the third molars, or wisdom teeth, developed as additional "grinders" to cope with cellulose-tough food. When wisdom teeth do come through today, they're often crowded or uncomfortably angled, and are frequently removed—you don't need them anymore.

DISSECT THAT SMELL

Many mammals, notably dogs, have a very highly developed sense of smell—so much so that they can single out components of a complex odor in the same way as humans can focus on small individual areas of color. And these mammals have an additional sensory organ, usually located just inside the nose and connected through an opening to the mouth, over which they can "run" the molecules of a smell to catch every particle of its significance. Called the vomeronasal organ (VMO), it plays a big part in the acuteness with which its owner experiences smells. When your dog looks as though he's running a particularly appealing scent over his palate, like an oenophile taking in a fine wine, he's probably using his VMO. While it's long been accepted that traces of a VMO—in the form of patches of extra sensory cells—can be found in humans, there's been plenty of academic debate on whether or not they are functional, or whether they're another example of an evolutionary leftover in the body.

The appendix, now a small, dead-end tube leading off the top of your large intestine, also had a role helping with the digestion of a plant-heavy diet. As the human diet changed, it shrunk and, most experts believe, became redundant (a few dissenters maintain that it may still play a not-yet-understood role supporting the immune system).

Guarding Vision

If you look at the inner corner of your eye, you'll see a tiny fold. It's all that remains of a third eyelid, the nictating membrane, that still exists in a number of other animals, from polar bears to sharks. A clear third eyelid, it could be pulled across the eye, keeping it moist and protecting it. It limited the need to blink; important when humans hunted a lot and relied on excellent vision and lightning reactions to succeed. Over time it was eventually reduced to just the trace visible today.

Are you really taller in the morning?

You are taller when you get up than you will be last thing at night, although not by much—if you measured yourself carefully, you'd find the difference would be between half to three-quarters of an inch. But why does it happen?

Forces of Gravity

It's the result of the effects of gravity on the soft matter in your spine. Although the bones of your spine won't compress, when you stand upright, the bouncy disks of cartilage between them will—and the same will be true of other "padding" in the body, such as the cartilage in your knees. When you're lying down at night, you gradually stretch out again, so by the following morning you're back at your tallest.

Taller in Space

Astronauts, measured after a stint in space with zero gravity, had "grown" more than the half-inch or so that you "shrink" after feeling the effects of a day's-worth of gravity while you stand upright. Some were as much as 2 inches taller than they had been before they went voyaging in space. The effect didn't last, though; once back in the world's atmosphere, over the course of a few months they returned to their original height.

YOU WEIGH LESS, TOO...

Weight-watchers know that the best news comes if you step on the scales first thing in the morning, when you'll find you're a pound or three lighter than you will be later in the day. The reason? Most people don't drink (or eat) during the night, they lose moisture in breath and sweat, and they urinate as soon as they get up—and the net result is a loss of (mostly) water weight. You're also burning energy in the form of calories while you're asleep. After you get up, your weight will gradually rise again as you eat and drink.

Which is your strongest muscle?

Although there are many extraordinary muscles in the human body, there's no one standout star, because they excel across a wide variety of functions, including strength, elasticity, and endurance.

High Performers

Your muscles divide into three groups: cardiac (your heart muscle), smooth (in your intestine and other organs), and skeletal (those muscles that are attached to, and work with, bones). Specific high performers include the masseter, in your jaw, used for chewing; the soleus, in your lower leg, crucial for flexible foot movement; and the gluteus maximus, the big muscle in your buttock, key in keeping you upright.

Unsung Heroes

There are also some quiet stalwarts. The minute muscles around your eye, for example, don't get much attention until they tire, but they are in a state of constant, delicate adjustment as you move from one task to another. And smooth muscles, which govern the workings of your intestines, operate quietly and automatically: You don't become conscious of them until something is wrong.

The heart, though it tends to be categorized as an organ, is all muscle, and perhaps wins the overall prize for sheer persistence. While most muscles are called into play as they're needed for different jobs, from running to eating, provided that you're alive, the heart is on call all day, every day.

TONGUE TWISTER

Despite popular opinion, your tongue is not your strongest muscle. The reason that so many people think that it is, is down to its stamina and flexibility. Actually, it's a fusion of eight separate muscles, unusual in that they aren't supported by any bony structure but twisted together into a form that's technically known as a muscular hydrostat (the same sort of muscular structure as you see in an octopus undulating across the ocean floor). However hard your tongue works, it doesn't tire, because if one muscle starts to flag another will take over the labor.

How many miles of tubing does the average body contain?

The lengths of some of your internal systems are astonishing. If you could take it out and uncoil it, your intestine would stretch out to between three and four times your own height, while your nervous system, if you did the same, would be over 45 miles long.

Stand-Out Record-Breaker

The most astounding statistic when it comes to the quantity of material that your body has packed into it is the overall length of your blood vessels. Taking the three types (arteries, which carry oxygenated blood from the heart, veins, which carry the "used" blood back, and capillaries, which link the two), the length of all the vessels ranges from around 60,000 miles in a child to over 100,000 miles in an adult.

How Far Could They Go?

100,000 miles is too big a number to be instantly relatable. But it's nearly half the distance between the Earth and the Moon—take three people, extract their sanguinary systems, unravel them, and you could use the arteries, veins, and capillaries to propel blood further than the Moon. As is the case with so many such statistics, this wouldn't be the best use of them. Far better that the incredible circulation system stays safely tucked up inside your body where it belongs.

NOT A HIT

The wonders of the human blood system haven't always impressed an audience. William Harvey, the physician and academic who discovered it and, in 1628, wrote about it, complained to John Aubrey, the biographer, that after he published his theory, he " . . . fell mightily in his Practize, and that t'was beleeved by the vulgar that he was crack-brained . . ."

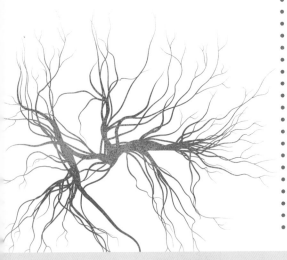

FASTEST, LONGEST, GREATEST, STRONGEST

The human body produces a good number of record-breakers. To test if you're the sharpest tool in the box, see if you can answer these quiz questions.

Questions

1. A sneeze goes faster than a speeding bullet—true or false?

2. Is the volume of tears you produce over your lifetime great enough for you to take a bath in it?

3. A hemispherectomy is an operation to remove half of which: your liver, your heart, or your brain?

4. Which part of your body did the Greek physician Galen claim worked the hardest?

5. Why might a dusty house offer health benefits?

6. What's a vomeronasal organ?

7. Would you be taller or shorter in outer space?

8. Which of your body systems, uncoiled, would reach from New York City to Westport, Connecticut, a distance of 45 miles?

9. Are dust mites insects or arachnids?

10. Why do you need eight different muscles in your tongue?

Turn to page 212 for the answers.

WHY ARE HUMANS BIGGER THAN THEY WERE 500 YEARS AGO?

HISTORICAL BODIES

DID CAVEMEN SUFFER FROM ALLERGIES?

DID HUMANS EVER HIBERNATE?

Why are humans bigger than they were 500 years ago?

It's common knowledge that humans are much larger and taller than they used to be. Like all "common knowledge," though, this doesn't tell the whole story. Our species' growth pattern is a story of fits and starts, rather than a steady progression.

How Big Did We Used to Be?

Looking at ancient suits of armor, you might conclude that humans have become a lot bigger. After all, there's not much chance of wriggling into a suit made from metal—it has to fit, and some look very small. But although humans are generally taller than they have been at some points in the past, the rate at which this change has occurred has been far from steady. Scientists have found that the size increase has happened erratically, and most growth graphs have some notable blips along the way.

Ups and Downs

Most accessible long-term studies deal with America and Northern Europe, so they're necessarily limited. One, published in 2017 by the University of Oxford, looked at the average height of men in England over a 2,000-year period. Its findings revealed that height went down in some periods as well as up, the determining factors including long-term weather conditions (for example, the "Little Ice Age" that began in the fourteenth century), periods of prosperity (ensuring a good diet, particularly important in childhood), and the different lifestyles of town and country dwellers. But both this and other broader studies had some surprises.

DEM BONES

How do you assess the height of someone when you have only their skeleton to go by? Forensics teams measure a long bone, usually the femur or thigh bone (although the long bone of the arm, the humerus, is also used), then apply an equation—for example, measuring the length of the femur in centimeters, then multiplying the measurement by 2.6 and adding sixty-five to get the height. Calculations vary a little according to the race and gender of the skeleton's original owner.

A 200-year Growth Spurt

Between the end of the medieval period and the start of the eighteenth century, people actually got shorter—by 1700, the average Northern European man was a full 2.5 inches smaller than he had been back in the eleventh century. The most consistent upward growth has come over the last two centuries; from the middle of the nineteenth century, Northern Europeans and Americans have become steadily taller, with North Americans holding the record as the tallest people in the world over the period between the American Revolution and the end of World War II. Since then, they've been overtaken by a number of European countries, with the Dutch currently holding the record; the average Dutchman measures up at 6 feet 0.3 inches.

Napoleon Complex

Final food for thought: The Napoleon Complex, said to afflict short men, is a modern misnomer. In fact, Napoleon measured up at 5 feet 6 inches—well above the average (around 5 feet 4 inches) for his time.

Did cavemen suffer from allergies?

id our ancestors, emerging from their caves and falling into fits of sneezing, discuss in grunts how it was obviously a high pollen-count day? Or are allergies a modern phenomenon, most likely attributable to our fussy, overly hygienic lifestyle?

It's in the Genes

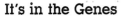

Homo sapiens overlapped (and interbred) with two other species of hominid, both long since extinct— the well-publicized Neanderthals and the less familiar Denisovans, a group living in Siberia about whom much less is known. Two studies published in the *American Journal of Human Genetics* in 2016 argued that this early interspecies breeding resulted in a useful genetic inheritance: We have inherited three TLR genes (TLR1, TLR6, and TLR10, to give them their full identification), whose job it is to see off disease, two derived from the Neanderthals and one from the Denisovans. And the effects of these don't seem to have diluted down thousands of years.

What is a TLR Gene?

"TLR" stands for toll-like receptor; these are the genes that can build the specialized proteins that ingest invading pathogens in the body and help to prevent infection. Since they had little in the way of effective medicine, it was particularly important for early humans that the body was able to stop infections developing. TLR genes aren't specific, as there are too many different potential pathogens for there to be a gene dedicated to each one. The proteins they make patrol the surfaces of the body's cells, searching for potential intruders. If they can't

deal with an unwelcome incomer themselves, they trigger the body's immune system response. And it seems that the TLR genes still work in very much the same way as they did around 50,000 years ago, when *Homo sapiens*, Neanderthals, and Denisovans were interbreeding.

Benefits with Baggage

This help in fighting off disease, though, came with some less-welcome side effects. Back when the three early hominid species were mingling, *Homo sapiens* was the incomer. The other two groups had been living in the same territories for millennia, and had adapted to their surroundings, making them immune to many of the pathogens in their habitat. The newcomers didn't have the same protection, but breeding with other hominids worked over time to increase their protection from disease. The downside was that it also increased their allergic response—meaning that their immune systems would sometimes act against non-harmful pathogens (think of the typical pollen allergy) and cause unwanted reactions.

THE HYGIENE HYPOTHESIS

What about the theory that cleanliness has caused the huge increase in allergies over the past 50 years? The idea was first raised in 1989 in a paper by Professor David Stachan, published in the *British Medical Journal*. With a growing, all-around emphasis on clean, bacteria-free surroundings, and an equally steep rise in allergies, particularly in children, he offered an argument that linked the two. Around three decades later, current thinking is that it's only one of several factors. It is thought that a baby born by caesarean section is more likely to develop allergies later in life, as is one that is formula- rather than breast-fed; both of these prevent a child from getting maximum benefit from the microbiome, the gut flora, of its mother.

What was the glass delusion?

If you've ever even thought about it, you might assume that madness doesn't change much with time. But there's evidence that some mental disorders belong to their era—after all, people couldn't fantasize about being radioactive before anyone knew that radiation existed. Perhaps the most extraordinary example of this is the glass delusion.

A Shattering Condition

Those who suffered from the condition believed they were made of glass. This gave them terrible problems. One of the first cases to be documented, at the

beginning of the fifteenth century, was King Charles VI of France. Handsome, magnetic, and initially a popular ruler, for the last 30 years of his reign he had lengthy periods when he believed his body was made of glass. He spent most of his time in bed, cushioned with thick layers of linen and wool; when he had to move, he wore a kind of corseted robe sewn with iron stays, to ensure that he wouldn't "break." Nor was he the only aristocrat to be thus afflicted. An even more convoluted case, and one of the last to be recorded, in the 1840s, was that of Princess Alexandra Amelie, daughter of Ludwig I of Bavaria, who maintained that in childhood she had swallowed a grand piano—but one made from glass. As a result, she moved gingerly through her parents' palaces, easing herself around corners, terrified that she would smash the glass piano into bits.

Although the records show that glass delusionists were wellborn and highly educated, this may simply be because sufferers from less eminent backgrounds were ignored, laughed at, or locked away (and not included on the records). For about two centuries, the glass delusion was a widespread problem, afflicting hundreds of people, widely recognized and written about, before gradually fading away again. The English academic Robert Burton includes it in his book, *The Anatomy of Melancholy*, published in 1621. "[They believe] that they are all glass, and therefore will suffer no man to come near them . . ." he wrote (although he also includes people who believe they are made from cork, and others who think they have frogs living inside them, so it's part of a fairly comprehensive list). It was a well enough known condition, too, to appear in Cervantes' 1613 story, *El Licenciado Vidriera* ("The Glass Graduate").

Madness Moves On

The glass delusion is almost unheard-of today. The first cases were noted at the beginning of the fifteenth century—when clear glass was a novelty. Professor Edward Shorter, a scholar of the history of medicine at the University of Toronto, has pointed out that, before the heyday of the glass delusion, some people believed that they were made from pottery. After the glass delusion ceased to be reported, in the nineteenth century, doctors wrote about obsessives who thought that they had turned into another newly-minted material, concrete. Could it be that some forms of mental disorder simply land on a scientific novelty and adapt themselves? After all, today, no one would be surprised to hear that someone suffering from a mental disorder believed that they were being influenced by hostile forces coming from their laptop.

66 [They believe] that they are all glass, and therefore will suffer no man to come near them . . . 99

Are humans really naked apes?

Well, it's certainly true that humans are apes—with the much-quoted fact that we share 98.8 percent of our DNA with chimpanzees, it would be hard to make a case against it. But if we're so closely related, why is it that humans, alone among primates, don't have fur?

Three Theories

All kinds of ideas have been put forward, although no single argument has yet won all-around agreement from both archeologists and paleontologists. There are three main threads of debate.

The first is that we were originally ocean-dwellers (see pages 66–67). This one argues that, at some time during their evolution, humans were at least partially aquatic. While there are fully ocean-going mammals, such as whales and dolphins, that don't have hair, seals —at home in and out of the water— do; semi-aquatic humans might have grown sleeker, but there's no convincing reason to think they would have lost their hair altogether.

Temperature Control

Perhaps the strongest argument that's been advanced for humans being the only non-hairy primates is the ability to sweat—and people sweat heavily. Sweating is the most efficient way to cool down, but it doesn't work fast

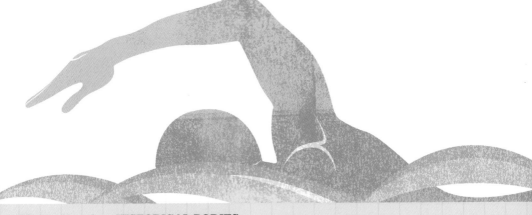

PLANET OF THE APES

If you believe in evolution, you already accept that you and all other humans are apes (or primates, perhaps—it sounds so much more civilized). But it's still hard to take in quite how closely we're related. Nonetheless, many abilities that are widely accepted as exclusively human are actually shared by other primates. Opposable thumbs? Monkeys and apes have them. Laughter? Chimpanzees and some other primates laugh if they're tickled (although, to be fair, it seems that rats do, too). Walking on two legs? There's evidence that some species of apes are comfortable walking bipedally at least some of the time. Tool use and making shelters are also primate skills, not just human ones. We may not be quite as exclusive as we think we are.

enough to be effective on skin that is thickly covered in hair. Humans seem to have begun to hunt on open grasslands around 3 million years ago, and unshaded, prairie-like conditions may have meant that they needed to shed hair and begin to sweat effectively. If this was the case, becoming "naked apes" would have been a practical evolutionary solution.

Pest Elimination

The third, recent, argument is that humans lost their hair as a way of ridding themselves of pests and sickness. Ticks and lice, and the—often serious—diseases they spread, don't settle on hairless skin; it's too hard to get purchase, and too easy for the target to brush them off. Once humans had smooth, naked hides, the theory goes, they were also largely relieved of the problems of many skin parasites.

Which, if any, of the three theories is the right one? While the intriguing aquatic idea has few supporters, each of the others has strong adherents. Although, of course, there's still the very real possibility that some altogether new idea may arrive to challenge or eclipse them.

Has anyone actually been raised by wolves?

From Romulus and Remus to Mowgli in *The Jungle Book*, there's no shortage of mythical or fictional children brought up by wild animals. But could it really happen, and has there ever been a proven case of someone who was raised by wolves?

Wolf Children or Tall Tales?

There are plenty of alleged cases of abandoned or feral children having being "raised" by other species; the difficulty has always been in establishing the factual evidence, which is often clouded by wishful thinking, folklore, or commercial considerations. One of the most fully documented cases was of a pair of Indian girls called Kamala and Amala, the so-called "wolf children of Godamuri," which dates from the 1920s. Despite a mass of both photographic and word-of-mouth evidence, it was ultimately suspected of having been a hoax perpetrated by the clergyman who "civilized" the children—like so many others, their story as told just had too many contradictions.

The Story of Ramu

More recently, in 1985, the tale of Ramu the wolf boy was reported by *The Times of India*, shortly after his death. Allegedly he was found, aged about two, in 1979, in the largely rural state of Uttar Pradesh, running on all fours with a family group of young wolves. He could not learn to speak, which is something true of many feral

children; it seems that the language impulse, if not prompted in infancy by the company of other humans, can be hard to trigger later on. Unlike many of the other stories, this one seemed hard to disprove, and it may be the closest we can come so far to a true raised-by-wolves case.

His later caregiver believed that his mother had left her child in the grass verges while she labored in the fields, and that a she-wolf, already rearing her own cubs, had come across the baby and simply added him to her brood. Ramu would never be able to tell the true story of his early years; like most other wolf-children he died young, aged around 10, still unable to talk.

FUSSY EATERS

You might wonder why a hungry wolf wouldn't gobble a human child up? Wouldn't a baby be more likely to make a delicious dinner than a play date for cubs?

While a very hungry wolf might eat a child, when it comes to hunting, wolves are actually rather conservative, contrary to their savage reputation. They stick to what they're used to and are cautious about new foods "on the hoof." So, a child might not immediately come to grief. Wolves, though, do eat most forms of carrion, so if a wolf found a human body, it would probably eat it. And having tried human, it might develop a taste for more.

When was smoking healthy?

The answer of course is never, but although it's hard to imagine now, there have been times in the history of smoking when tobacco was regarded as an aid to health, although there had been warning signs that it could harm you quite soon after it arrived in Europe.

First Opposition

Thomas Harriot, a friend of Sir Walter Raleigh, and the first to import tobacco to England, died of mouth cancer in 1621. As early as 1604, James I issued his *Counterblaste to Tobacco*, in which he decried the habit of smoking as "harmful for brains and dangerous for lungs." But it was such a profitable import that ultimately the king simply raised a massive tax on it and kept the use of tobacco legal.

Cure Your Cough with a Cigarette

From the mid-nineteenth century, medical links between cancer, heart disease, and smoking were becoming increasingly clear, but advice often emphasized that it was smoking too much—rather than smoking at all—that was bad for you. Advertising recommended smoking to cure tickly coughs. And by the early twentieth century, smoking had come to be perceived as glamorous, too. As late as 1910, Dr. George Meylan of Columbia University could write, " . . . there is no scientific evidence that the moderate use of tobacco by healthy mature men produces any beneficial or injurious physical effect that can be measured" (smoking was thus neither good nor bad for you, though note the words "healthy mature men").

Endorsed by Doctors

In the 1940s, nervous pregnant women were still being told by doctors that smoking would make them less anxious. Medical periodicals such as the *Journal of the American Medical Association* still carried pages of tobacco advertising—some featuring white-coated doctors puffing away: "More doctors smoke Camels than any other cigarette!"

Through the 1950s and 1960s, though, smoking became officially bad for you. In January 1964, the U.S. surgeon general held a press conference in which he stated unequivocally that smoking caused lung cancer. Finally the battle lines against smoking had been drawn: You might choose to smoke, but not even the tobacco industry could pretend any longer that it was a harmless habit.

GLOW-IN-THE-DARK GOOD

Tobacco isn't the only substance that has been erroneously believed to be healthy. At the turn of the twentieth century there was a craze for radioactivity in all kinds of products. A contemporary article in the *American Journal of Clinical Medicine* raved about it in terms that are hard to credit today. "Radioactivity prevents insanity," it crowed, "rouses noble emotions, retards old age, and creates a splendid, joyous, youthful life." Radithor, an expensive radiated water, sold briskly to those who wanted to enjoy all these wonderful effects. One enthusiast, the prominent U.S. industrialist Ebenezer Byers, drank it regularly from 1927, when it was "prescribed" to him as a painkiller to deal with an intransigent arm injury. The effects would kill him in 1932, a few months after Radithor was finally taken off the market in the face of increasingly worrying research. The *Wall Street Journal's* epitaph on Byers was unsparing: "The radium water worked fine," it read, "until his jaw came off."

Whatever happened to the body of Thomas Paine?

Today there are usually two mainstream options when someone dies: Their body is either buried or cremated, and both processes are subject to quite a few regulations. In the past, though, there weren't as many rules.

A Memorial Back Home

In the case of the philosopher and revolutionary Thomas Paine, his body would become the subject of gossip, myth, legend, and folklore—long after

the remains themselves went missing. By the time of his death in 1809, the great pamphleteer had been living in the United States, in poverty and political unpopularity, for some years. He had asked to be interred in the Quaker graveyard at New Rochelle in New York, but the Quakers rejected him, and eventually he was buried in the grounds of his own farm there. Back in England, where he was born, a few of Paine's supporters considered that this wasn't good enough for a father of the revolution—they felt that he should be brought back to his native land and a collection made to raise funds for the memorial.

Restless Bones

William Cobbett, journalist and friend, did more than discuss the idea; he acted on it. A decade after Paine's death, Cobbett sailed to New York, dug up the coffin, bagged the body and returned to England with it. His escapade was greeted with a mixture of disapproval and mockery. Even Lord Byron got in on the act, penning a sarcastic verse. Such was Paine's posthumous unpopularity that back in England Cobbett found that he could raise no interest in a subscription for a memorial, or even a memorial dinner. Eventually, Cobbett cut locks of hair from the rotting skull and used them to fashion a number of memorial rings, but even these found few takers. Finally, the bones went into a box and up to Cobbett's attic, where they were stored until his own death in 1835. When the contents of the house went on sale, the bones narrowly escaped being auctioned. Instead, they were passed to another friend of Paine, Benjamin Tilly, who hung onto them until he died himself in 1869. And then the trail goes cold, or at least lukewarm. Various people subsequently claimed ownership of Paine's bones, or at least some of them, but they have never been run to ground.

FOR THE REVOLUTION

One of the strangest rumors that sprang up around the saga of the body of Thomas Paine was that some of his bones had been cut up and made into buttons to be sold for the revolutionary cause. It's almost certainly an apocryphal story, although it's possible that some buttons were sold under such pretenses. Even at the time, contemporary wits said that there were far too many of them: at least ten bodies' worth.

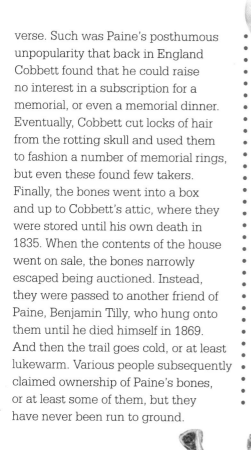

Were humans originally aquatic?

Of all the theories around the process of evolution, possibly the one that has raised the most debate is the question of whether humans ever lived in water. First raised in the 1960s, the idea is largely dismissed by serious historians today, but it still attracts plenty of enthusiasts and regular arguments break out as to its validity.

Where it Began

It started small—in 1960, Sir Alister Hardy, an eminent marine biologist, published an article in *New Scientist*, "Was man more aquatic in the past?" It outlined the basic ideas of what was later to become known as the aquatic ape theory, and was given a brief mention in *The Naked Ape*, a popular study of human evolution by Desmond Morris, published in 1967. Elaine Morgan, originally a scriptwriter, who authored a number of popular science books, then took up the argument with the publication of *The Aquatic Ape,* published in 1982.

Taking Off

Hardy's original proposal was very modest, with plenty of qualifiers. He did suggest that a number of features of *Homo sapiens* had a possible affinity with a life in, if not under, water— humans' hairlessness, smooth bodies, unusual degree of breath control, and thick layer of subcutaneous fat were all cited. As time passed, however, plenty of arguments against it were raised, and number one among them was the absence of any fossil record. More and more fossils of early hominids were being found in land that had historically been forest and grassland that didn't offer any potential watery habitat.

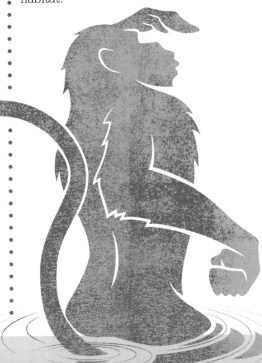

Appealing . . . But Wrong

Another problem arose when anthropologists tried to fill in the holes in the aquatic ape theory. The idea of early people living in water was satisfyingly simple, but the evolutionary adaptations that seemed at first to support it could be proved to have happened over many millions of years, covering periods and territories where a sea- or riverside life wouldn't have been possible.

A Middle Way

In 2009 Richard Wrangham, a primatologist at Harvard, suggested that, rather than being largely or semi-aquatic, early humans had led a waterside lifestyle, using water plants and roots (which were more likely to be available year-round than land plants) to fill in nutrition gaps at lean times of year. The fact that apes living in grasslands that regularly flood still have a tendency to walk upright through water because it's easier to balance—chimpanzees regularly wade in flooded rivers in the Congo, for example—was used to support the argument. Could it be that our ancestors, rather than being fully fledged divers, actually led a wading lifestyle? It's still unproven, but the aquatic ape theory shows no signs of going away just yet.

JAWS

Aquatic ape theory supporters have also claimed that water would have made a safer habitat than land, enabling hominids to avoid the predations of, say, big cats. Not so, retort the naysayers: Crocodiles were numerous in the water, and would have been just as dangerous.

Did humans ever hibernate?

Humans were hairless from quite early on, so how did they survive cold winters? Hibernation seems like an obvious answer. You eat heartily while there is still food around, then go to sleep for the coldest months, slowing down your metabolism dramatically. In spring you wake, hungry but alert, ready to go out hunting again.

Staying Warm

The scientific consensus, however, says no. There's nothing in the way in which humans evolved that indicates we ever hibernated. In general our resistance to very low temperatures is poor: While a dormouse, say, can survive sustained body temperatures as low as 30.2°F, humans tend to have trouble when their core body temperatures fall just 3 or 4 degrees below their 98.6°F standard and remain there. Also, humans only began to leave the relatively warm climates in which they evolved, traveling to colder ones, around 100,000 years ago, by which time they had certainly already discovered the uses of fire. A hundred millennia sounds like a long time to us, but it's a short hop in evolutionary terms—not long enough for humans to have become hibernators, particularly if there was a warm hearth to come home to.

HYPOTHERMIA MIGHT BE GOOD FOR YOU

Despite the absence of evidence for ancestral hibernation, the idea of a kind of induced hibernation is gaining interest. Many surgeons have already adopted the practice of cooling patients down for complicated surgery, lowering the heartbeat and slowing the metabolism—thus buying extra time and reducing potential trauma. Some experts think that the next step will be quite a few degrees colder, putting patients into a sort of suspended animation.

HISTORICAL BODIES

Even though our bodies always work in the same way, past perspectives on them often differed considerably. See how much you've learned about our ancestors with this quiz.

Questions

1. Is the Napoleon complex said to afflict people who are a) Corsican, or b) short?

2. Which early hominid hailed from Siberia?

3. Wolves are fussy eaters—true or false?

4. What is the peculiar use to which the bones of the journalist and political activist Thomas Paine are said to have been put?

5. Name two species, apart from humans, who laugh if they're tickled.

6. Which king of France believed he was made from glass?

7. In which Indian state was Ramu the "Wolf Boy" first discovered?

8. James I of England was a strong advocate of smoking tobacco—true or false?

9. Is radiated water good for you?

10. Give one reason why archaeologists think humans were never aquatic.

Turn to page 213 for the answers.

HAS WHITE ALWAYS BEEN BEST WHEN IT COMES TO TEETH?

IS THERE AN IDEAL LENGTH FOR EYELASHES?

COULD A BODY BE 100 PERCENT COVERED IN TATTOOS?

FASHIONABLE BODIES

Could a body be 100 percent covered in tattoos?

Over the last 40 years or so, tattoos have moved from mainly nautical territory into the mainstream. A survey carried out in 2017 estimated that 40 percent of U.S. adults aged between eighteen and sixty-nine have tattoos. They're . . . ordinary.

Pricey and Painful

Some tattoo enthusiasts may have a full sleeve or two of designs. But what about the extreme enthusiasts who want to go *much* further than a few discreet inkings? If you wanted to, would it be possible to get the whole of your body tattooed?

First, it would cost a lot. Most tattooists charge by the hour, and, depending on how virtuoso and how dense you want your patterns to be, body inking can take a very long time. Star tattooists are busy, and may not accept all comers. The pain would be a consideration, too. Most describe it as irritation rather than agony, but if you wanted your whole body covered, even with the simplest designs, you could expect high-level discomfort at the very least—more so in your more sensitive corners.

BODY MODDING

Even extreme tattooing sits at the more familiar end of the body modification market. Real body modification enthusiasts—body modders—may go much further, with extreme piercings, tongues cut into forks, implants under the skin to give the impression of horns or scales, or scarification, in which the skin is cut to create patterns of raised scars.

The Difficult Bits

What about the body parts that aren't just skin? Fingernails, for example, or eyeballs? Fingernails aren't hard to tattoo, but they grow out, so they don't last long—every time you trim them, you'll be taking a bit of your new nail art with the clipping.

Those who've had it done describe the sensation of needles pricking into your nails as being spectacularly irritating, too; much worse that the stinging-nettle sensation of a skin tattoo.

Your tongue? Eurgh. But yes, you can have it tattooed, although the uneven surface may lead to a slightly crude effect, and the average tongue gets such heavy use that the pattern will wear off faster than on some other parts of your body.

What about eyes? You can have your scleras, the whites of your eyes, colored, but even the tattoo artist who popularized the idea of scleral tattoos in 2007—Luna Cobra, a peripatetic modifier who travels all over the world—agrees that it isn't a safe thing to do. The operator injects color into the eye, but only allows the needle to go to a very shallow depth, just under the conjunctiva, the protective covering of the eyeball. Then they "flood" the area with the chosen color. You can't have a design tattooed on your eyeball, so the technique is really more akin to dyeing. It can be a risky procedure: Too much ink may mean that color travels to other areas of the face, leading to ink stains where you don't want them, and can also result in bad headaches, increased light sensitivity, temporary blindness, and some very nasty infections. In conclusion, yes, you could get your whole body tattooed, but it would be expensive, painful, and potentially risky.

When was plastic surgery invented?

Today we tend to think of plastic surgery as a modern art; the immediate connection you might make with the term would be "improvement" surgery, such as breast or buttock enhancement, or facelifts. However, it was first developed in ancient times to repair wound damage or the ravages of disease.

First Days

The first mention of plastic surgery comes in the *Sushruta Samhita*, by the Indian healer Sushruta, written in the sixth century BCE. Among a good deal of information about healing plants and the treatment of diseases, he described both skin grafting (the technique of "patching" wounds with skin taken from other parts of the body) and surgery to reconstruct the nose— the first rhinoplasty, in which a flap of skin from the forehead was used to graft onto a damaged nose.

Seven hundred years later, the Roman physician Aulus Cornelius Celsus covered grafting and reconstruction in *De Medicina*, the surviving fragment of a much larger work.

Step-by-Step Surgery

Perhaps the first pioneering European surgeon was Antonio Branca, himself the son of a Sicilian surgeon, who is credited with inventing the "Italian method" of skin grafting and rhinoplasty.

A German, Heinrich von Pfolspeundt, recorded Branca's method in fascinating detail in his *Buch der Bündth-Ertznei*, published in 1460. First the surgeon made a kind of template for the nose in either parchment or leather; this was laid down on the patient's forearm and traced around. The shape was cut around to create a flap of skin, but left attached to the forearm at the lower edge, the forearm raised to the face, and the free portion of the flap stitched in place over the nose.

The graft was then left for 10 days to "take," the patient spending the time with their arm strapped to their face so the graft could not be pulled away. After this time, the lower flap was cut through, so that the arm could be freed and the final, lower part of the graft stitched in place around the nostrils. The huge advance in Branca's method was that it ensured the skin was "alive" by leaving it attached to its original site while the graft took place—although it must have been very stressful for the patient.

WHY "PLASTIC" SURGERY?

The term was invented by Pierre-Joseph Desault, a surgeon and teaching professor at La Charité and Hôtel-Dieu hospitals in Paris in the second half of the eighteenth century. He was the first to use the term *plastikos*, from the Greek, meaning "able to be molded," in connection with the operations, usually skin grafts, that aimed to correct or improve scars and facial deformities. In allowing large numbers of students to attend operations, he created a new generation of surgeons who followed his innovative methods and techniques.

Has white always been best when it comes to teeth?

We brush, we floss, and we bleach, all with the intention of keeping our teeth as clean and white as possible. But has white always been the preferred color for teeth, or have there been times and places when other hues have been fashionable?

The Practice of *Ohaguro*

While there's no record of yellow teeth ever having been much admired in any society, black teeth have enjoyed periods of popularity, particularly in the Far East. Laos and Vietnam have both had traditions of dyeing the teeth black, and in Japan, *ohaguro*, as the practice is called, was popular for centuries. Black teeth, contrasted with stark white face makeup, became fashionable at some point during the Heian period (between the eighth and twelfth centuries), and carried on all the way until the end of the Edo era at the close of the nineteenth century. White teeth simply weren't admired, and writers compared catching a flash of white inside someone's mouth to glimpsing a mouthful of worms: very unalluring. Initially, *ohaguro* was mainly practiced in the upper echelons of society, but

over the years it filtered down through the social classes, although it remained more popular with women than men.

Fashionably Black

In Japanese culture, rich black was traditionally highly admired and considered exceptionally beautiful,

so pure black teeth carried the same value that pure white teeth carry today. The effort put into making and drinking the dye, too, was no greater than that taken to get today's desired bright-white effect—which can involve long sessions of bleaching, or even applying veneers.

How to Do It

Teeth were first prepared by being rubbed with pomegranate rinds, which apparently made it easier for the black color to "take." The dye was made from a mix of tea and *kanemizu*, a liquid made from iron filings. It was drunk as soon as the mixture turned black and, in turn, it would blacken the teeth. It needed to be taken almost daily for the effect to persist, though, and the metallic-tasting mixture was flavored with spices to make it less disagreeable to drink.

In 1873, when the empress of Japan appeared in public with naturally white teeth, she caused a considerable stir. The empress was supporting a movement to open up Japan to outside influences and release it from the isolation of previous centuries—as

part of the modernizing process, *ohaguro* had actually been forbidden 3 years earlier, in 1870, but the fashion took time to die out. Today, blackened teeth are seen only occasionally, usually on a geisha or maiko (apprentice geisha).

FASHIONABLE TOOTH DECAY

Black teeth were also briefly fashionable in England in Tudor times. Imported sugar became available in the course of the sixteenth century, but it was prohibitively expensive. Since it also rotted teeth, and rot turned one's teeth black, blackened teeth, the thinking went, demonstrated one's wealth and privilege. Records don't reveal what the rotten-toothed fashionista's solution was for the resulting bad breath.

Why did men stop wearing high heels?

Perhaps the real question is, why did men start wearing high heels in the first place? The original reason was practical: When they were on horseback, the heels stopped riders' feet sliding out of the stirrups—the first heels for men were on the boots of Persian cavalry soldiers.

Worn in the Saddle

The Persian army was known for its mounted archers, who needed to stand up in the stirrups to shoot efficiently, and the heels enabled them to do this without slipping. When, at the end of the sixteenth century, the Persian leader Shah Abbas sent a delegation to Europe to ask for help in defeating the increasingly powerful Ottoman Empire, the novel and exotic dress of the delegates (and their heeled footwear) attracted attention, and elements of their costume quickly became modish at the European courts. Initially, there was no hint of the feminine about these early high heels—their connection to the military meant that the message they sent was masculine, even martial.

Badges of Rank

Of course, out of the saddle, high heels weren't necessarily practical to walk in, but in a sense that was the purpose of them: Aristocratic men didn't need to do manual work, and an elegant high heel proved the point. They went on to become popular with royalty; both Charles II of England and Louis XIV

Perhaps the last refuge of the macho high heel is found in the cowboy boot. Like the boots of the original Persian archers, the heel is intended to help you keep your foot in the stirrup, and can be anything up to 2 inches high, although on a classic cowboy boot it will always be a slanted Cuban heel.

of France wore extravagantly elegant and expensive versions. Louis XIV, at a merely average—for his time—5 feet 4 inches, added an impressive 4 inches with lavishly decorated boots and shoes. His footwear always had bright red heels and soles, and he even introduced a rule that only courtiers could copy the color. (Nearly four centuries later, an echo of this exclusivity would come when designer Christian Louboutin defended in court his singular right to the red soles that he had made his trademark as a shoe designer.)

An End to Heels

Although the men led the way, women soon followed, and high heels became universal among the noble classes in Europe until, as always, the fashion wheel shifted and clothing, particularly for men, became plain and sober with the arrival of the Age of Reason in the seventeenth century. Embroidery, bright colors, and elaborate fabrics were all discarded, and by the mid-eighteenth century, men's shoes were once again flat and unornamented. Women's shoes also became comparatively low, modest, and plain.

High heels would once more become fashionable in the mid-nineteenth century, but heels for men would never again achieve the same level of open popularity—although sneaky "elevators" would be placed invisibly in shoes to give short men a valued additional inch or two, and in the 1970s high platforms for both sexes enjoyed a brief year or two in the fashion sun.

What was the point of foot binding?

In China, the practice of foot binding for women lasted for almost ten centuries—well beyond the usual lifetime of most mere fashions. Today it's often viewed with horror, as it left women's feet deformed and near-useless. But what was it actually for?

Golden Lotus Feet

The Chinese legend often quoted to explain the origins of "lotus" (bound) feet tells how one of the favored concubines of the tenth-century Emperor Li Yu had danced for him, perched on a golden lotus pillar, her tiny feet bound with ribbons. This, probably apocryphal, story was used to explain the custom whereby girls' feet were bound to keep them small.

The Binding Process

For a "successful"—minute—result, feet had to be constricted when the girl was very young, sometimes only 3 years old, while they were still soft enough to be molded to shape. The feet were first soaked, then folded, so that the heel and the big toe met, losing the natural arch. The other four toes were tucked tightly underneath the sole of the foot, so that the end result was a triangular shape, pointing directly downward from the ankle.

Women interviewed in the late twentieth century remembered that the binding process was extremely painful for the first year or two, but that after this time, their feet became numb and settled into their constricted shape. Of course, women subjected to this process couldn't walk easily or far. Bound feet in their naked state were not pretty to look at, but they were shod in highly decorative, often exquisitely embroidered shoes (whole factories were dedicated to their production), to set off their dainty size.

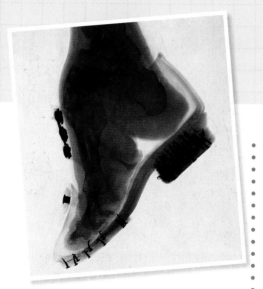

Erotic Fetish, or Practical Planning?

The traditional view of bound feet was that they were both a status symbol—an economic "boast" that a girl did not need to work—and a beauty fetish, aimed at making girls more appealing and marriageable. Over time the custom moved outside the aristocracy and became widespread across different social classes, although it remained exclusively practiced by the Han Chinese.

Authored by anthropologists at McGill and Central Michigan Universities, a book published in 2017, *Bound Feet, Young Hands*, posed an altogether different view from the long-held idea that foot binding showed that a girl did not need to work. It argued that a good deal of the economic work of communities in China was done by women at home—they wove and spun, sewed and knotted fishing nets, all key to supporting a family. And bound feet, by making walking very difficult for them, ensured that they would be bound to their work—at home. When such work began to be done in factories, having women tied to the home no longer made economic sense, and the practice of foot binding died out.

SMALL, SMALLER, SMALLEST

Tiny bound feet were judged according to the lotus scale, with the smallest feet gaining their owner the most kudos. Smallest of all were "golden lotus" feet, just 3 inches long. At 4 inches, they became "silver lotus" feet, while—comparatively—large, though still objectively tiny, 5-inch feet fell into the less prestigious "iron lotus" category.

Is there an ideal length for eyelashes?

The hair on your head, if it isn't cut, will usually keep on growing—at least to shoulder length, if not longer. So what decides the length of other hair on your body? Eyelashes, for example, never seem to grow beyond a certain point—why?

The Perfect Third

Apart from being considered an aesthetically appealing frame to the eye, eyelashes also have a practical function. They keep dust and other irritants away from the eye's sensitive surface and control the flow of air around the eyeball, helping to ensure that it doesn't lose moisture and dry out. In 2015, a study done at the Georgia Institute of Technology set out to discover if there was an optimum length for eyelashes to perform these tasks most effectively. After experimenting in a wind tunnel, it was concluded that the ideal length for the mammalian eyelash (that's for every mammal, not just humans) was one-third of the overall length of the eye. There wouldn't be any point in lashes growing longer than this, as it would ultimately interfere with their performance. Makeup, particularly mascara that dramatically extends the length of the lash, is therefore not particularly good for your eyes, although the same tests done on longer, mascaraed lashes concluded that some protection around the eye, even if it didn't offer the best airflow control, was better than none.

THE LIFESPAN OF HAIR

As to why the hair on your head can grow longer than that anywhere else on your body—it's because the hairs on different sites have varying lifespans. While all hair goes through the same three stages—active growth (technically called the anagen phase); stagnation, when growth stops (the catagen phase); and drop (the telogen phase, when, after a period of dormancy, the hair falls out)—it does so on different timetables. Your head hair can grow for between 5 and 6 years before "stagnating" (the time that your hair is neither growing, nor falling out, but simply being there), while your underarm or pubic hair (and, in the case of men, chest hair, too) is on a much shorter cycle of, at most, 45 days.

It's in the Follicles

Every hair on your body grows out of a follicle, and you are born with the finite number of follicles you will have all your life. Follicles may go dormant, causing hair thinning and loss as they close up shop, but it takes a lot actually to kill them. In hereditary baldness, the follicles shrink and become shallower, and the hair they produce is sparser and finer. Eventually the follicles close up altogether. However, while many people lose the hair on their heads, shedding eyelashes or eyebrows is much rarer, and is usually the result of a specific condition, alopecia, which triggers the body to attack its own hair follicles.

Was there ever a practical reason for men shaving their beards?

The average man has around 25,000 follicles on his chin and around his jaw, so his potential for beard-growing is strong. Shaving is laborious and, before the arrival of really sharp razors, must have been quite painful. So where did the idea first come from?

Clean-shaven

Homo sapiens started early; he was using sharp shells to cut his beard before he had even discovered metal. The Ancient Egyptians shaved with copper razors, and Alexander the Great insisted his soldiers shave off their beards, according to Plutarch, on the grounds that beards were too easy for an opponent to grab in the heat of battle. But other civilizations valued their beards: The Assyrians, as their sculpture attests, had magnificent, long, curly beards.

Possibly men first shaved with an eye toward cleanliness. When it isn't possible to wash, dense, curly hair may be perceived as offering a potential home for all kinds of nasties—in World War I, most soldiers went rapidly clean-shaven, not because of regulations, but because of the plague-levels of nits and lice in the trenches.

Could Beards Harbor the New Penicillin?

In 2016, however, a study published by the alarmingly named *Journal of Hospital Infection* tested the bacteria of more than 400 subjects—all hospital workers—both bearded and beardless. And the results vindicated beard-wearers. The beardless carried higher quantities of harmful bacteria on their skin than their bearded colleagues. What's more, some of the microbes combed from the beards showed promising signs of antibacterial activity. At a time when known antibiotics are beginning to fail, perhaps the next generation will be discovered in a beard.

What's the most dangerous makeup ever worn?

Every beauty standard, in whatever era, seems always to have begun with smooth, blemish-free skin. And some very dangerous methods indeed have been used to get the look.

Beauty Kills

The worst-ingredients hit parade is probably topped by lead. It was ubiquitous in cosmetics from Ancient Egyptian times until the late eighteenth century, and often used as a thick mask, so the skin could hardly help but absorb it. Among other ghastly side effects, it caused sores and scarring, and users often applied an even thicker layer to cover them up.

Second in line comes arsenic, popular for ensuring a fashionably pale appearance in the Victorian era. (It was everywhere in the nineteenth century, so even if you avoided it in your face cream, you were quite likely to encounter it in your dress fabric or your wallpaper.) Side effects? Exhaustion, fever and joint pain, along with swollen hands and feet.

Third comes mercury—good for fading spots and freckles, and for smoothing out wrinkles. As well as smoother skin, it could also cause numbness, spotty vision, depression, and permanent damage to the kidneys and nervous system.

CHECK THE INGREDIENTS

You may be congratulating yourself that today's stringent health and safety rules must ensure modern makeup is completely safe. Not necessarily—in 2007, of more than thirty lipsticks tested in an independent laboratory in the United States, a third were found to contain unsafe levels of lead. And a number of ingredients that you might have assumed long-banned still make occasional appearances in cosmetic horror stories—mercury, for example, is still sometimes used for its preservative properties.

Do piercings have health benefits?

Piercings, like tattoos, have gone mainstream over the last three decades: They've moved from conventional earlobes-only to piercings of the septum, lip, tongue, and other areas NSFW. Besides sending social signals, though, can piercings have any practical use?

Pressure Points

In Ayurvedic medicine, specific piercings are believed to help with various health problems. Many correspond with the pressure points of Chinese acupressure and acupuncture, tapping in to the energy pathways in the body, to maximize their efficiency. The traditional earlobe piercing, though, carried out on small babies, sometimes at just a few days old, is probably judged the most important. The *Sushruta Samhita*, an encyclopedia of health written in the sixth century BCE, sets out the rituals for the *Karnavedha*, the double ear-piercing ceremony, which is one of the sixteen *sanskaras*—a complicated word which means something like "rites of passage." It also lists all the good things it will do for a child's health. Read through them, and you'll see why, as an all-around protective health

measure, the *Karnavedha* is key. Boys have their right ears pierced first, then the left; for girls, it's the other way around.

Know Your Meridians

Earlobes are thought to hold important meridian points that link with both sides of the brain. It's held that piercing them will help a baby's brain to develop properly and aid with memory function all through life. There is also a point in the middle of the lobe which links with the reproductive system, and, for girls, an accurate piercing will ensure that their menstrual cycles stay healthy. Yet another point aids digestive function.

FASHIONABLE BODIES

There's nothing new about humans dressing up or decorating their bodies—it's been going on since prehistoric times. Take this quiz and find out how much you've absorbed about *Homo sapiens'* fashionable habits.

Questions

1. Why was it shocking when the empress of Japan appeared in public with naturally white teeth in 1873?

2. What is scleral tattooing?

3. What were men's high heels originally useful for?

4. Golden lotus, iron lotus: which was more prestigious in Ancient China?

5. Makeup often contained arsenic in the Victorian era. Can you name two other everyday sources of arsenic in the nineteenth century?

6. Alexander the Great made his soldiers shave off their beards because they were a battle hazard—true or false?

7. Who came up with the name "plastic" surgery?

8. What job do your eyelashes do?

9. The Ancient Egyptians used sharp seashells to shave with—true or false?

10. What kind of heel would you find on a classic cowboy boot?

Turn to page 213 for the answers.

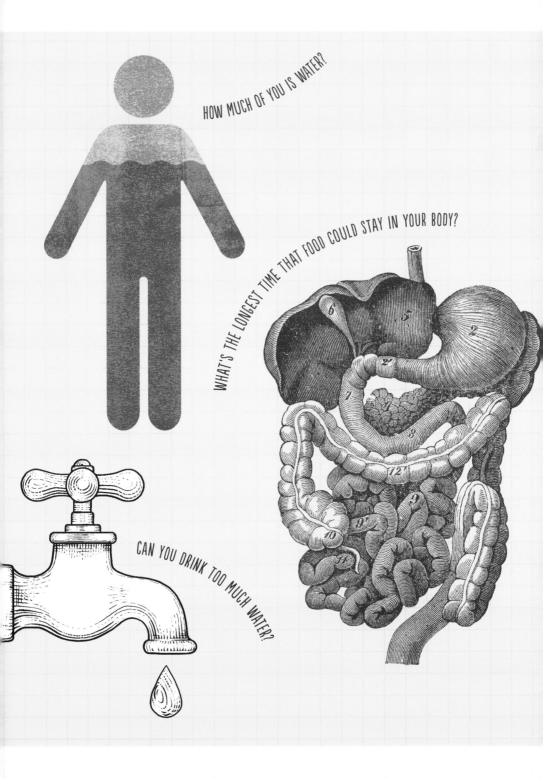

HOW MUCH OF YOU IS WATER?

WHAT'S THE LONGEST TIME THAT FOOD COULD STAY IN YOUR BODY?

CAN YOU DRINK TOO MUCH WATER?

WHY DOES YOUR STOMACH GROWL WHEN YOU'RE HUNGRY?

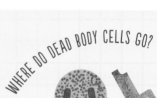

WHERE DO DEAD BODY CELLS GO?

INTERNAL AFFAIRS

COULD YOUR GUT CAUSE DEPRESSION?

IF 98.6°F IS YOUR NATURAL BODY TEMPERATURE, WHY DOES IT FEEL SO HOT OUTSIDE?

Can you drink too much water?

We're so used to being told that we need to rehydrate that the idea of overdoing it doesn't tend to cross our minds. But while it's very rare, it is possible to drink too much water—and, confusingly, the symptoms of over-hydrating, or hyponatremia, literally water intoxication—are very similar to those of heat exhaustion.

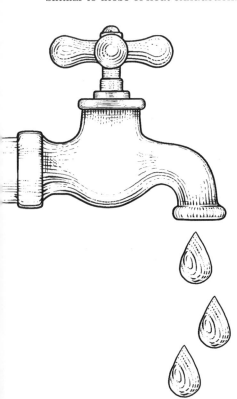

Drunk with Water?

In past times, most people just drank when they were thirsty. Today, there's less trust in such a simple solution, and recommendations come thick and fast: You need 4 (or 6) pints of water a day; you should drink eight glassfuls a day, you need 7 pints in addition to the water in the foods you eat, and so on. And of course, sometimes your body will need extra water. If you exercise heavily, you'll need to drink more to replace the amount you're sweating off—the same goes for hot weather. If you're pregnant or breastfeeding, you'll also need to drink more.

What Happens in Water Intoxication

All the same, if you drink too much too fast (and in cases of hyponatremia, this tends to mean gallons rather than glassfuls), your kidneys can't process the overload, your sodium levels drop, and your body starts to retain water. The side effects are nasty: Hands and feet will swell up; you may feel sick, and actually vomit; you're likely to develop a deep, thumping headache because your brain is swelling; and you may have a sudden bout of diarrhea. Left untreated, the pressure

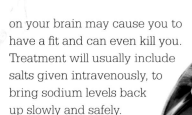

on your brain may cause you to have a fit and can even kill you. Treatment will usually include salts given intravenously, to bring sodium levels back up slowly and safely.

Bottling It

Exactly when did bottled water become so ubiquitous—even fashionable—in Europe and America? Although it was popular through the 1800s, the market languished after tap water became safe to drink in the early twentieth century. Its renaissance in the United States began when Perrier, until then a relatively obscure brand of French sparkling water, launched a major marketing campaign in 1977, leading with a series of modishly shot television ads, with voiceovers supplied by the gravelly tones of U.S. actor Orson Welles.

It was a successful move—by 1979, Perrier was selling 200 million bottles a year. Other water companies started to take notice and to put substantial marketing budgets behind their own brands. By the early 2000s, a bottle of water was as necessary an accessory as a pair of Juicy Couture velour joggers. And the fetishizing of bottled water seems to be here to stay—in 2017 annual sales of bottled water in the United States overtook those of all other drinks for the first time. In the future the horrendous environmental impact of all those plastic bottles may downgrade its popularity, but for the time being, water rules— so long as it comes out of a bottle, and not the tap.

Why does your stomach growl when you're hungry?

It's a familiar feeling: An hour or three after you ate, you're somewhere quiet—a meeting, a play, your best friend's wedding service—and your stomach starts up, grumbling and growling as though you'd swallowed a disgruntled bear. Why the noise? Is it really a sign of hunger? What does it mean?

Digestion Live

Actually, the growling noise is as likely to be originating from your gut as your stomach, and it isn't necessarily happening because you're hungry. It also goes by the much grander onomatopoeic (that is, the word is intended to imitate the sound of the process it describes) name of borborygmus, and has its origins in the Ancient Greek word for "rumbling." What you're actually hearing is your digestion hard at work.

The rumbling is the sound of food, liquid, and gas being churned around by peristalsis, the squeezing, muscular action that propels the contents of the stomach into, and then through, the gut. (Think of a soft mass, combined with quite a lot of air, being squeezed around in a constricted space.) It's a noisier process as the stomach gradually empties because there's more space for pockets of gas and air, and it's these pockets that make the noise. So the growling is most likely to happen a few hours after you've eaten—the food and drink are broken down by digestive juices and move from the stomach to the small intestine and ultimately the colon.

THE SOUND OF SILENCE

Your gut may be pretty active much of the time, but the digestive processes slow down at night. Even if the stomach is relatively empty it's unusual to hear loud gurglings when its owner is deeply asleep.

RUMBLE

RUMBLE

RUMBLE

Where Does the Air Come From?

You swallow quite a lot of air as you eat (one of the more practical reasons you were probably told not to talk with your mouth full as a child). What's more, your swallowing reflex becomes less efficient as you get older, and you tend to take even more air in. Not only that, but the digestive processes produce gases as by-products: carbon dioxide, hydrogen sulfide, and methane, to name just three. With what you swallow and what you make, your digestive system is a highly gassy environment.

What Happens Next?

By the time it leaves the stomach for the small intestine, your last meal is a mass of partially digested matter known as chyme. That's far from the end of the story, though.

The muscular walls of your stomach and small intestine contain sensory receptors which tell your brain when it's empty and which trigger waves of electrical activity. These, known as migrating motor complexes, first prompt the beginning of hunger pangs for your next meal—which will gradually strengthen until you become conscious of the fact that you're ready to eat again—and second, cause a further wave of contractions from the stomach into the small intestine, which clears any debris from your last meal, leaving the way open for the whole process to start over again when you next eat.

If 98.6°F is your natural body temperature, why does it feel so hot outside?

People differ in their responses to warm weather, but most would agree that when the thermometer hits 98.6°F on a tropically sunny day, they feel hot—even uncomfortably hot. But why, when that's the exact internal temperature of the human body?

Heating Up

Whether you're resting up or moving around, your body is in a constant state of activity, from your beating heart to your firing brain synapses. Even if you were the laziest couch potato in the world, numerous processes would still be carrying on automatically, and every metabolic process creates heat. So it's important for your body to be able to cool down efficiently.

A Built-in Thermostat

The hypothalamus, a small area at the base of your brain, is charged with regulating your body temperature. It helps to even up the heat created in the different areas of the body so that you're a more-or-less even temperature all over most of the time. But when it comes to getting rid of excess heat into your outside surroundings you are still dependent, to an extent, on the temperature of your habitat.

When you're cold, it's the hypothalamus that sends the direction out for you to start shivering, putting the body into motion to warm you up; when you're hot, on the other hand, it will direct your body to sweat and to increase the amount of blood flowing through the vessels closest to the skin's surface, to give your body the best chance of cooling down.

IT'S NOT THE HEAT, IT'S THE HUMIDITY

It's the catchphrase that gets repeated over and over again in muggy, sticky weather. But why do humidity levels make such an unwelcome difference to the way you experience the heat? It's because humidity affects your ability to sweat effectively. Usually, when you break a sweat in the heat you cool down as the liquid evaporates off your skin. But if you're sweating in air that is already saturated with moisture, the sweat won't evaporate, and you're left feeling hot, damp, and uncomfortable.

Baby, It's Warm Inside

It's a rule of physics that when two objects of different temperatures are brought together, the one with the higher temperature will transfer heat to the one with the lower temperature. This is why a cup of hot coffee left out on the kitchen counter will gradually cool to room temperature. In the same way, your body will leak heat out into cooler surroundings. If the external temperature becomes equal to, or even higher than, your internal one, the heat has nowhere to go, so you will find it increasingly difficult to cool down.

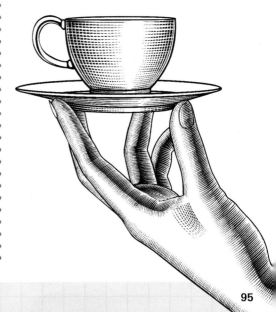

How much of you is water?

Adult humans are generally made up of as much as 65 percent water, while the level is even higher in children (65–68 percent). Highest of all is the percentage in small babies, who may consist of as much as 78 percent water. That's only 18 percent less than a cucumber.

The Crucial Element

Generally women have a lower water content than men, because they tend to have a greater percentage of fat, which contains less water than leaner muscle tissue. And as you age, your water percentage shrinks and you dry out slightly: Very old people may go down to as low as 50 percent water.

So where is all that water hiding, and what jobs does it do? Looking at organs and tissues one by one, the amount of water you contain might not be what you expect. Blood is liquid, so surely it must have the most water? Wrong, though it's true that plasma is mostly water, at over 90 percent. But plasma makes up only just over half of your blood, so the end result is that blood doesn't come top of the water-content list.

One of the first analyses of the exact chemical contents of different parts of the human body appeared back in 1945 in the United States, in the *Journal of Biological Chemistry*. It found that the wettest of your body parts are the lungs (around 83 percent water), followed by the muscles and kidneys (approximately 79 percent) and the brain and heart (around 73 percent each). Your skin is around 64 percent, and even your bones are almost a third water. Over 70 years later, these conclusions are still accepted as broadly accurate.

Working Water

The water inside you isn't just floating around: Two-thirds of it is confined within individual cells. Some obvious water-related jobs are the regulation of body temperature (sweating) and getting rid of body waste and toxins (urinating and defecating), but other roles include acting as a cushion and shock absorber for the joints, brain, and spinal cord, and serving as a crucial building block in cell generation. Water also carries the nutrients you need to function around the body as the blood circulates.

WHAT WOULD HAPPEN TO YOUR BODY WITHOUT WATER?

If you've never experienced any water deprivation greater than a slight, easily slaked degree of thirst, here's the sequence of (grisly) events. How long it will take depends on the external environment, but it isn't pretty. First, your kidneys won't send any more water to your bladder. Then, you'll stop sweating—you'll feel hot—and the increased concentration of your blood will make it flow sluggishly—so you'll feel faint. At the next stage, your organs will start to fail, as the blood is no longer reaching them. In the final stages of dehydration, your body will no longer be able to control its temperature, and your kidneys and liver will cease to function. Shortly after that you will die. Stay hydrated!

Where do dead body cells go?

We're accustomed to hearing (and being revolted by) the fact that the house dust that accumulates around our homes is largely made up of human skin cells that have been sloughed off in the course of day-to-day activity. But what about the cells inside the body? How long do they live, and where do they go when they die?

How Cells Die

Cells in your body die at a high rate. Scientists estimate around 50 billion perish in a single day. They can die in two ways—by apoptosis or necrosis. Most types of cell have a fairly set life span, and apoptosis is what might be called a cell's "natural" death: When it has outlived its usefulness, it begins to dismantle itself from within. Proteins inside the cells called caspases prompt the production of enzymes that destroy the DNA of the cell and start to break it down. As part of the process, the cell becomes "leaky"; it sends out messages to the specialist cleaner cells, called phagocytes, which will deal with the debris of the dead cell.

Death by necrosis is less well regulated. It happens because a cell has been damaged in the course of a trauma to the body, such as an external injury or an infection. The sudden death isn't necessarily neat and self-contained like apoptosis. The dying cell, which bursts rather than leaks, won't send out the same signals as an apoptotic cell, so it may not be so easy for the phagocytes to detect and remove it efficiently after it dies—although they will still "collect" a necrotic cell for breakdown. The chemicals released by necrotic cells can also trigger areas of inflammation in the body.

WHEN GOOD CELLS GO BAD

Sometimes the DNA of cells isn't disposed of properly—this can happen when necrotic cells die, and also sometimes if the phagocytes have had too much work to do. If this happens, it may provoke the autoimmune system of the body to overreact, in some cases causing serious conditions, such as lupus, anemia, and arthritis.

The Body's Rubbish Collectors

The phagocytes that help to clear up after dead cells are two kinds of specialist white blood cells: neutrophils and macrophages. Although they're often referred to as being the rubbish collectors of the human body, their role is really more akin to that of very efficient recyclers: They literally engulf the remains of a dead cell and help to degrade them so that their component parts can be reused by the body. They're produced in your bone marrow, then travel in your blood to where they're needed. When a dead or dying cell is carried into their orbit, they'll move swiftly to gobble it up.

What's the longest time that food could stay in your body?

Despite a much-increased general interest in human digestive processes over the last couple of decades, there's still a lot of confusion over how long food takes to digest (and perhaps even more over how long it should take). So what's the "healthy" time frame for food to make its way through your body, from your mouth to the toilet?

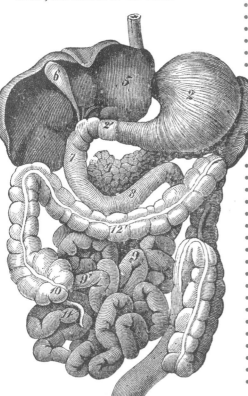

It Depends

It's one of the most frustrating answers, but in the case of digestion rates, it's true. The conclusion from a wide variety of studies into digestion is that there's a large personal variation. One of the most widely publicized and enduring studies was made in the 1980s by the Mayo Clinic, which found that on average the whole process takes a little over 50 hours, and that food spends the vast majority of that time processing through the large intestine, having sped through the stomach and the small intestine in a relatively rapid 6 to 8 hours. Women generally have slower digestion than men, and certain foods take longer to process. Subsequent studies have brought the timing down, in some cases quite considerably—some have shown a total digestion time of something closer to 30 hours.

THE DIGESTION/ ABSORPTION MIX-UP

It's common for people to confuse the digestive process—the journey of your food from your mouth to the point at which it's expelled as waste—with absorption, the process during which your body absorbs the good stuff in your food and converts it into the forms in which it can use it. Most absorption happens in the small intestine between 2 and 7 hours after eating. The small intestine is lined with thin-walled protrusions called villi, which increase the surface area of its tubing massively. As food is digested, the molecules pass through the villi into the bloodstream. From this point, they can be transported around the body to where they're needed. By the time the residual food passes into the large intestine, the nutrients have been taken out of it and absorption is over. The job of the large intestine is to remove the water from the remaining mulch, after which what's left is feces, ready for excretion.

So How Long Should Digestion Take?

There isn't a "best" time. Although certain conditions, such as Crohn's or other types of inflammatory bowel disease, can either speed up or slow down your digestion, this is an area in which there is no real normal. People eat a very wide variety of foods, and the rates at which different foods are digested varies, too. If your digestion doesn't give you any pain, you don't suffer from constipation or diarrhea, and your bathroom visits are more or less regular, you probably don't give your digestive system much thought. And, provided that you eat reasonably well, ideally following the U.S. author and journalist Michael Pollan's mantra, "Eat food. Not too much. Mostly plants," you're probably fine. Or, to take another, much older, mantra: "If it ain't broke, don't fix it."

Could your gut cause depression?

Around a decade ago, if you'd asked a doctor what caused depression, you probably would have been told that it was due to a chemical imbalance in the brain; specifically, low levels of a neurotransmitter called serotonin.

The Brain-gut Connection

In recent years, though, fresh examination of what goes on in your gut has rewritten the script. The view that's gaining scientific traction is that the gut microbiome—the huge number of bacteria, good and bad, that colonize your gut—may be the starting point for all kinds of things that were previously held to be the remit of your brain.

The surprising thing about this discovery is that anyone is really surprised. After all, we've always known that emotional upheaval can cause stomach upsets, and expressions linking mood and stomach are common currency: You may be "sick with nerves," for example. So why shouldn't the brain-gut link go further than that? And could it go the other way, too, with gut problems causing trouble for the brain?

Of Mice and Men

So far, most of the major studies into a link between depression and imbalances in the microbiome of the gut have been done on rodents. They've established that low levels in certain key bacteria elicit something that looks very like depression, with an increased readiness to stop trying in challenging situations. Of course, people aren't mice, but extensive studies of the influence of gut flora on mood in humans are in the pipeline. The new science, christened psychomicrobiotics, which looks at potential medical uses for all kinds of microorganisms that occur naturally in our guts, is definitely in the ascendant.

Is there an optimum time of day to eat?

As a child you were probably told not to skip breakfast, and it's likely that you heard that it was best to "Breakfast like a king, lunch like a prince, and dine like a pauper." Does when you eat your food make any difference to your health?

Breakfast Like a King

The thinking behind the saying was that it's best to eat the most at the time when you need energy, rather than waiting until you're naturally winding down. Eating a large meal late in the day can mean that your body stores calories as fat, rather than using them as energy.

The Eating Window

Recent advice has changed the reasoning a little, and defeated the argument that you should eat little and often. A study at the Salk Institute for Biological Studies in San Diego showed that mice that were given the opportunity to eat across only 8 hours in every 24 were healthier than mice given the same rations, but allowed to eat whenever they wanted. A British study took this as a starting point for an experiment with human subjects. They were asked to eat the same foods, in the same quantities, as usual, but to eat breakfast 90 minutes later than usual, and dinner 90 minutes earlier, adding 3 hours to their natural overnight fast time. After 10 weeks, the "fasting" group had lowered their cholesterol and blood-sugar levels, and many had also lost weight, while the control group's levels and weight had stayed the same. So if you want to be healthier, it seems as though it may be worth looking at when, as well as what, you eat.

How many types of bacteria lurk in your gut?

We hear a lot about the importance of the microbiome in your gut. The last decade has seen an explosion of interest in the workings of the human digestive system, and new studies have made the bacteria in your colon answerable for everything, from stress levels to intelligence.

Gut Interests

You certainly have plenty of them—trillions of bacteria, in fact, weighing in at an estimated 2 to 4 pounds per person. They are highly individual; most people harbor between 500 and 1,000 types, both beneficial and not so good, and everyone has a slightly different formula.

Do You Have a Gut Type?

Most people know their blood type, but do you know what type of gut you have? In 2011, a study in Heidelberg, Germany, found that there are three broad profiles of gut microbiome. This didn't mean that guts of the same profile had identical microbiomes, but that they shared most of the same bacterial groups, and missed others altogether. The people who shared each profile seemed to have nothing else specifically in common. They didn't share ethnicity, and they were a mix of genders, and various ages, weights, and states of health.

The Good, Bad, and Useful

The identification of gut microbiome profiles is probably just the beginning. Scientists hope that in the future it will be possible to tailor medical treatment according to the "formula" of a patient's gut, adding microbial reinforcements to the bacteria already there. With antibiotic resistance becoming increasingly and worryingly common, it could be that in a decade or two, antibiotics may be replaced by using bacterial adjustments to an individual's gut to improve their immunity to disease.

INTERNAL AFFAIRS

We all know about the bits of the body we can see—hair, skin, fingers, toes, and so on—but what about the bits we can't? How much have you learned about them?

Questions

1. When you drink too much water, is the resulting condition called hydrophobia or hyponatremia?

2. How much of the water in your body is confined within your cells?

3. Can you name three gases produced as by-products of the digestive process?

4. Which contains the highest percentage of water, a newborn baby or a cucumber?

5. Necrosis or apoptosis: which name describes the "natural death" of a cell?

6. What weight of bacteria are you carrying around in your colon?

7. Your blood is the body part that contains the most water—true or false?

8. Villi are a) mini air sacs within your lungs, or b) small protrusions that line the walls of your small intestine?

9. What does the new science of psychomicrobiotics study?

10. Which is healthier when it comes to meals—a) little and often, or b) consumed inside a shortish time window within your day?

Turn to page 214 for the answers.

WHAT'S THE BEST WAY TO SURVIVE A LIGHTNING STRIKE?

COULD YOU LIVE ON A VAMPIRE DIET?

UNEXPECTED EVENTS

COULD A REGULAR PERSON SURVIVE A ZOMBIE APOCALYPSE?

Can fear really turn your hair white?

It's a common horror-movie scenario: Something so traumatic happens that it causes someone's hair to turn white—overnight. It's become known as Marie Antoinette Syndrome, after the ill-fated French queen whose hair supposedly went pure white the night before she was sent to the guillotine. But could it really happen?

All In the Roots

The hard science says no. The hair on your head is dead matter, so once it's grown, it can't change color naturally. The color comes from two types of melanin, cells for both of which, called melanocytes, are in the follicle, close to the skin's surface. These two—pheomelanin, which adds the red or blond shades, and eumelanin, which produces the darker brunette and black colors—between them make up all possible variations of human hair color. As you get older, the follicles gradually cease to produce melanin, and the hairs growing out of your head will lose their color, one by one. What we call gray hair is usually a mixture of pigmented and unpigmented hair (melanin-free hair is actually colorless; a whole head of melanin-free hair will appear white, though, due to reflected light).

Any color change has to come from the root, and will only become visible as the hair grows out of the follicle. *Canities subita*, the technical term for the rapid whitening of the hair, would take far longer than 24 hours, say experts—realistically, months rather than days.

Alternative Explanations

"And in the morning, when they unlocked the door and let him out, his hair had turned pure white…" makes a great ending for a ghost story, but if it can't be true, is there any alternative explanation? In 2013, the *International Journal of Trichology* made a study of unusually rapid hair-whitening. Taking 196 cases reported from 1800 onward, they found forty-four that could be authenticated, and sought some realistic explanation of the phenomenon. None was found that convincingly fit the circumstances.

Two possibilities were offered, however: The first was that subjects could be suffering from *alopecia areata*, an autoimmune reaction that causes patches of your hair to fall out suddenly. If, by some quirk, only the pigmented hair fell out, it could appear as though

THE WHITE QUEEN

As for Marie Antoinette, there may be an altogether simpler answer to the question of her suddenly white hair. French artist Jacques-Louis David's sketch of her en route to the guillotine shows her in a simple shift, the ends of her hair poking rather pathetically out from under her pleated cap. Historians have suggested that maybe her jailers denied her either a wig or a more concealing cap—revealing for the first time hair that had turned white over a long period of sorrow.

someone's hair had gone white overnight. But the hair would inevitably have thinned considerably—and in none of the cases was any such thinning reported. The second idea was that, in theory at least, it would be possible to wash artificial color out of dyed hair to render it white overnight. Despite best efforts, neither option seemed completely satisfactory.

Could you live on a vampire diet?

Dracula and a whole clan of Cullens did it—although of course they're fictional. If you wanted to, though, would it be possible to live on an all-blood diet?

Any Old Iron, and Other Concerns

The difficulty of working out the pros and cons of vampirism is that they're highly hypothetical. In the absence of genuine vampires, any facts and figures have to be speculative. The potential pros (you wouldn't want for protein) are wiped out by some pretty hefty cons (if you discount iron, blood contains negligible amounts of many vitamins and minerals). But if you were hell-bent on going Gothic in this particular way, here are some of the other things that you'd be up against.

You'd need to put away quite a lot to meet your energy needs. It's estimated that human blood contains 430–450 calories per pint (the amount of a typical blood donation). So an adult male would need to drink about 6 pints per day to get his 2,500 calories. Women could manage on slightly less.

Blood is very high in iron, but your system—which can tolerate up to 45 milligrams daily—could probably cope (although if you did overdose, it could cause a life-threatening condition called hemochromatosis, which leads to a whole raft of problems, including

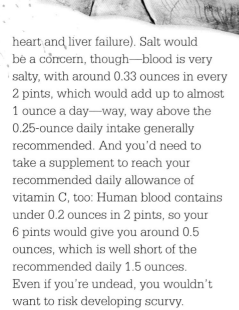

heart and liver failure). Salt would be a concern, though—blood is very salty, with around 0.33 ounces in every 2 pints, which would add up to almost 1 ounce a day—way, way above the 0.25-ounce daily intake generally recommended. And you'd need to take a supplement to reach your recommended daily allowance of vitamin C, too: Human blood contains under 0.2 ounces in 2 pints, so your 6 pints would give you around 0.5 ounces, which is well short of the recommended daily 1.5 ounces. Even if you're undead, you wouldn't want to risk developing scurvy.

On the Plus Side

With their routine of sleeping through the daylight hours, there's one aspect of vampires' lifestyle that is right up to date. Humans in the twenty-first century are often told that they're chronically short of sleep, and current thinking is that you need at least 8 hours a night—possibly as many as 10. So the vampire average of at least 10 hours a day (naturally they'll be awake through the night) would be one healthy habit you could emulate.

BETTER FOR BATS

Just three mammal species are known to live on blood alone—and all three are different kinds of vampire bat (as well as the common variety, there's also a hairy-legged vampire bat and a white-winged one). Unlike humans, vampire bats have evolved to cope with such a specialist diet. They boast super-sharp teeth to cut through skin, anti-coagulant saliva—thanks to the presence of a glycoprotein with the wonderful name of Draculin—to allow their victim's blood to flow freely enough for them to lap it up, and an exceptionally unusual gut microbiome, which, among other things, features around 280 bacteria that would cause serious illness in other animals.

QUESTION 50 QUESTION

Do human beings really spontaneously combust?

Cases have been recorded since the seventeenth century, and it's been a popular topic in fiction, too. But is spontaneous human combustion really a "thing," or is it a fantastical take on tragic but perfectly explicable deaths by fire?

How is it Defined?

Spontaneous human combustion has always made the headlines. Both Dickens, who made it a fitting end for the gruesome rag-and-bottle peddler Krook in *Bleak House*, and the Russian author Gogol, who used it to finish off one of the titular characters in his glum classic *Dead Souls*, are said to have used contemporary newspaper reports as sources. Most alleged real cases have certain factors in common: The victims were on their own when they died; although their bodies have been reduced to ashes by intense heat, their extremities—usually lower legs or feet—may survive the fire intact; their surroundings are untouched by the flames; and there's no indication of how the fire started. With just the right mixture of horror and mystery, spontaneous combustion has scored innumerable top spots on "unexplained"-style shows.

The Wick Effect

If you leave aside the question of how a fire started, the peculiar way in which the bodies are consumed can be satisfactorily explained by something

> **❝** I'm left with the conclusion that this fits in the category of spontaneous human combustion, for which there is no adequate explanation. **❞**

called the wick effect. Human bodies are fatty (and the majority of victims are heavy and sedentary, so may be judged fattier than average). If a piece of clothing or lock of hair catches fire, it's believed that it can act as a "wick," burning down a person in the same way as the wick of a candle, with the subcutaneous fat of the body acting as the "candle wax." This can lead to an intense heat consuming the body almost completely from within—which goes some way to explaining why the surroundings of many victims aren't damaged. A well-known experiment carried out in 1998 by Dr. John de Haan sought to replicate the process by wrapping a dead pig in a blanket, then soaking a corner of the blanket with paraffin and setting it alight. The results looked very like most known cases of spontaneous combustion: The pig was burned to ashes (apart from its trotters), but the room it was in suffered only minimal smoke damage.

How Does the Fire Start?

The source of the fire, then, is probably the most debated aspect of these cases. In 2011, Dr. Ciaran McLoughlin, a coroner in Ireland, recorded the first-ever official verdict of spontaneous human combustion on the death of Michael Faherty, who died at home from extremely severe burns, leaving his surroundings completely undamaged. "I'm left with the conclusion," he said, "that this fits in the category of spontaneous human combustion, for which there is no adequate explanation." The reasoning behind the verdict was not so much the intensity of the fire as the fact that no source for the fire could be found. Believers in spontaneous combustion offer various arguments for fire breaking out inside the victim's bodies, from a buildup of methane to some as yet unexplained internal cellular activity. The coroner may have given his conclusions, but further afield, the jury remains out on the verdict.

Why can amputees still feel their missing limbs?

Ambroise Pavé, a notable French surgeon, was the first to write about phantom limb syndrome in the mid-sixteenth century. It's the phenomenon by which amputees feel a nonexistent limb is still there, and is sometimes painful, too. So what's going on?

The Ghost in the Brain

Phantom limbs occur in a very high percentage of amputees—some studies have estimated as high as 95 percent. The sensation that the missing limb is

still present is so powerful that a subject may even move as though their leg or arm is there: If a ball is kicked toward them, for example, they'll move to kick it back, attempting to use the limb that isn't there.

There are many theories about where phantom limb syndrome originates, from the nerve endings near the amputation to messages within the spinal cord. The majority of scientists believe that it goes back to the brain and to the somatosensory cortex, which takes in and deals with information gained by touch. A number of studies have found that the brains of many amputees don't seem to have registered that a limb was missing: The area of the brain that would have been responsible for moving and controlling it had often hardly altered at all. And it's possible that this "things as normal" situation in the brain causes the sensation of a phantom limb.

The Mirror Treatment

For a long time, the main treatment for phantom limb pain was conventional painkillers, which seemed to have very hit-and-miss results. But in the early 1990s, the neuroscientist V. S.

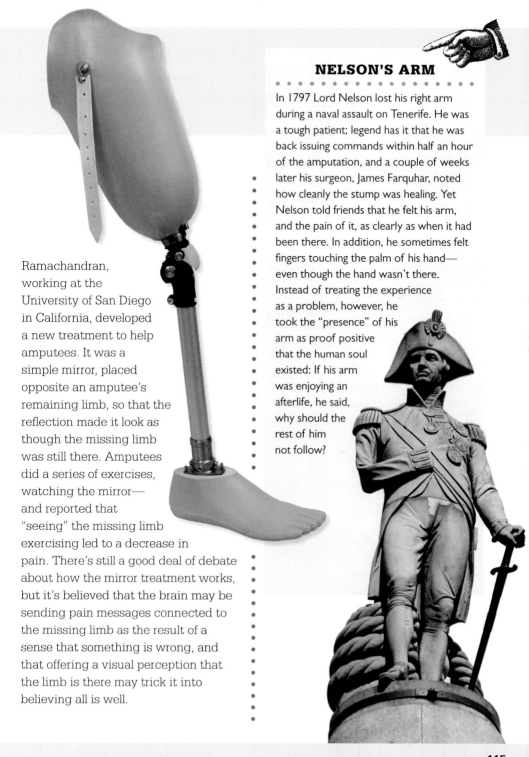

Ramachandran, working at the University of San Diego in California, developed a new treatment to help amputees. It was a simple mirror, placed opposite an amputee's remaining limb, so that the reflection made it look as though the missing limb was still there. Amputees did a series of exercises, watching the mirror— and reported that "seeing" the missing limb exercising led to a decrease in pain. There's still a good deal of debate about how the mirror treatment works, but it's believed that the brain may be sending pain messages connected to the missing limb as the result of a sense that something is wrong, and that offering a visual perception that the limb is there may trick it into believing all is well.

NELSON'S ARM

In 1797 Lord Nelson lost his right arm during a naval assault on Tenerife. He was a tough patient; legend has it that he was back issuing commands within half an hour of the amputation, and a couple of weeks later his surgeon, James Farquhar, noted how cleanly the stump was healing. Yet Nelson told friends that he felt his arm, and the pain of it, as clearly as when it had been there. In addition, he sometimes felt fingers touching the palm of his hand— even though the hand wasn't there. Instead of treating the experience as a problem, however, he took the "presence" of his arm as proof positive that the human soul existed: If his arm was enjoying an afterlife, he said, why should the rest of him not follow?

Could a regular person survive a zombie apocalypse?

In 2016, a group of students in the Department of Physics and Astronomy at the University of Leicester, United Kingdom, undertook a meticulous study of what their chances would be in the event of a zombie apocalypse. Their findings were alarming—if you believe in zombies, that is.

Adapt to Survive

The assignment was an end-of-year fixture on a very creative course—students are challenged to apply their practical knowledge of physics in a hypothetical situation and write up the results in a paper published in the peer-reviewed *Journal of Physics Special Topics*. In the end, the students of 2016 submitted two papers. Perhaps the results of the first were just too grim? The students began with a classic SIR model, a standard method for estimating the spread of an infectious disease such as Ebola

over the course of an epidemic. (S is the number of susceptible individuals, I is the number of infected individuals, and R is the number of recovered individuals; in the zombie version, it became SZD—susceptible, zombie, dead.) In their first exercise, they decided that humans would not be able to fight back, and made the assumptions that every zombie would be able to locate a human victim each day, and that the victim would have a nine in ten chance of themselves turning zombie.

The theoretical zombie takeover starts quite slowly—Chris Davies, one of the contributors to the study said, "I thought it would be interesting to see . . . if the television shows portray the spread at an accurate rate

mathematically. There's not much activity for the first 20 days, but then there's a sharp spike and . . . not many people left at Day 100." The team ended up estimating that it would take just 100 days for zombies to wipe out most regular humans (final toll: just 273 surviving humans, each outnumbered by zombies by a million to one).

Fighting Back

The second paper introduced some more hopeful aspects (if you're a human, that is). It gave humans more ability to escape the zombies, and calculated that the latter would have only a 20-day lifespan if they couldn't access human prey. Not only that, but it assumed that people who could bear children would be doing so as quickly as possible. It was enough to tip the balance: At the end of the second paper, the calculations showed that the human race would come through.

BUT SERIOUSLY...

Given that the chances of zombie attack seem to be quite slight, what's the benefit of calculating our chances in the event of something that probably won't happen? Despite the not-so-serious direction of the Leicester study, it still offers an idea of how specialists in disease epidemics—epidemiologists—approach an outbreak of serious disease and the ways in which such an epidemic might be tackled. In a real epidemic, tracking the outbreak back to the first case is crucially important, as it enables specialists to track back and work out the basic reproduction ratio—that is, how fast and by what means the disease is spread. Once both speed and means are established—and only then—does it become possible to devise an effective strategy to contain and deal with the outbreak, whether it's a case of bad winter flu or zombies.

Is stress ever good for you?

We're accustomed to being told how stressed modern life renders us—which comes with the assumption that stress is necessarily a bad thing. But are there any circumstances in which stress might be good for you?

Good Stress, Bad Stress

Contrary to what lifestyle magazines might tell you, high levels of stress aren't new: Humans have been living stressful lives since the first *Homo sapiens* had to face down a Neanderthal as they competed for the last local mammoth. But the word itself is used as a catchall that can mean anything from a slightly over-busy work schedule or social diary to dealing with serious illness or death. To distinguish between good and bad stress, scientists tend to refer to "eustress" (good)—the kind of stress that provides the impetus to stretch yourself and make an effort—and chronic stress (bad), which acts as a long-term drag on your mood and health.

In humans, stress sets up an early warning that is detected by the hypothalamus at the base of the brain, which in turn prompts the production of a number of hormones, including adrenaline and cortisol. You've probably found that in a suddenly stressful situation you think more clearly and are more decisive and this is because of the call-to-action effect of freshly released hormones. Adrenaline (also called epinephrine) is raising your heart rate and blood pressure, while cortisol is prompting the release of more glucose into your bloodstream and simultaneously slowing down body functions that don't need your attention in an emergency. Not only that, but stress can send your immune system onto high alert, ready to defend you if need be. A slight degree of stress will motivate you, encourage you to focus on your immediate situation, and energize you—all things that also tend to make you feel upbeat and positive.

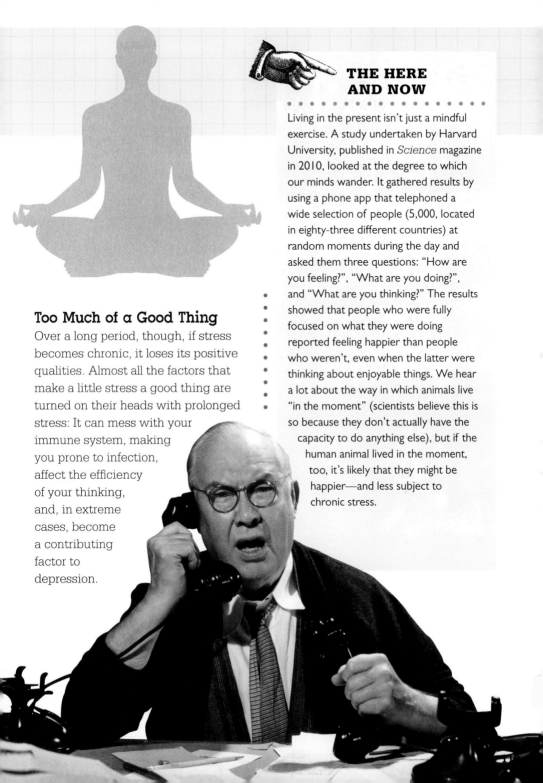

Living in the present isn't just a mindful exercise. A study undertaken by Harvard University, published in *Science* magazine in 2010, looked at the degree to which our minds wander. It gathered results by using a phone app that telephoned a wide selection of people (5,000, located in eighty-three different countries) at random moments during the day and asked them three questions: "How are you feeling?", "What are you doing?", and "What are you thinking?" The results showed that people who were fully focused on what they were doing reported feeling happier than people who weren't, even when the latter were thinking about enjoyable things. We hear a lot about the way in which animals live "in the moment" (scientists believe this is so because they don't actually have the capacity to do anything else), but if the human animal lived in the moment, too, it's likely that they might be happier—and less subject to chronic stress.

Too Much of a Good Thing

Over a long period, though, if stress becomes chronic, it loses its positive qualities. Almost all the factors that make a little stress a good thing are turned on their heads with prolonged stress: It can mess with your immune system, making you prone to infection, affect the efficiency of your thinking, and, in extreme cases, become a contributing factor to depression.

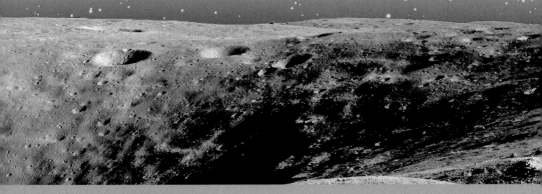

Does the human body work normally in space?

For the sake of the question, we'll assume that in this instance "in space" means either safely confined to a space station, or performing a short, timed space walk in a fully equipped space suit. If you were simply pitched into space without the suit, your end would be nasty, though quick. (More on that in the box opposite.)

Balance, Bones, and Growing Taller

Space travel messes with the delicate balance of your inner ear, so astronauts suffer from at least a day or two of space nausea. Most say it's no worse than other kinds of travel sickness. Weightlessness is reportedly harder to get used to; many astronauts find it difficult and uncomfortable sleeping in space (even when they're strapped in

place, their arms will float up and their heads will fall forward), while back home, the worst difficulty is that it takes a while to remember that you can't simply let go of objects mid-air. A longer-term problem is a loss of bone density, due to the absence of gravity—over a 3-month stay in space astronauts may lose at least 14 percent bone density, and there are records of some losing up to 30 percent. No one wants prematurely aged bones, so astronauts need to do at least 2 hours' exercise a day, to avoid losing either too much muscle or bone; they also go into a rehabilitation program of physiotherapy once they're back on Earth. Plus, it's recommended that they eat between twenty-five and thirty prunes every day—not to keep them regular (there are few reports of constipation in

If you were to float out into space without your space suit, you would have immediate problems, although you might survive for a minute or two. There is no air to breathe in space, so once you ran out of oxygen, you would lose consciousness. And attempting to delay this by holding your breath wouldn't be advisable, either: Any air retained would quickly expand and burst your lungs because of the lack of external pressure. While you were trying to solve your breathing problem, the liquid in your body would start to boil—this is known as bullism—because the vacuum of space would push the boiling point of fluids below your own body temperature. Back in 1966, an astronaut named Jim LeBlanc sprang a leak in his space suit during preflight simulated space conditions. He was rescued, but recalled that his last sensation before he lost consciousness was the feeling of saliva boiling on his tongue.

space) but because the very high levels of antioxidants they contain will also help to maintain healthy bones.

It's annoying, though not dangerous, that without gravity, the liquid in the body will rise—on Earth, it's gravity that helps to push fluids into your lower body. As a result, astronauts tend to have chipmunk cheeks (and blocked noses). They get taller in space, too, by around 3 percent, but in most people this adds up to a couple of inches at most (see page 46 for a full explanation for why this happens). An allowance is tailored into astronauts' space suits, so even at their tallest they will still fit into them. They will shrink back to their normal height once they return to Earth.

QUESTION 55

What's the best way to survive a lightning strike?

The best way to survive a lightning strike is the most obvious one: Avoid being struck by lightning in the first place. Statistically, more people are struck while fishing than doing anything else, so if you're a keen angler, you might want to reconsider your choice of leisure activity.

Worst-Case Scenario

Most people know the basics—try not to be caught out in the open during a thunderstorm. If you're within sound of the storm, you're in the risk area and are within reach of a lightning strike. Get onto low ground if you can, and make sure you're not near any tall trees. Crouch down on the ground, balancing on the balls of your feet, with your heels held together. Make sure your body isn't touching anything that could act as a conductor (a bag with metal clasps, for example). That way, you're making yourself as short as you can, with the smallest area possible in contact with the ground.

Am I Safe Indoors?

Not necessarily. You're certainly safer indoors than out, but avoid using any electrical appliances, and steer clear of water—it's not the moment to be taking a shower or bath. And avoid touching anything metal (window or door frames, for example) until the storm's over.

> **❝ You're certainly safer indoors than out, but avoid using any electrical appliances, and steer clear of water. ❞**

WORLD'S WORST

If you can stop worrying long enough, spare a thought for the people living in Kifuka, a village in the Democratic Republic of Congo, which has more lightning strikes annually than anywhere else on Earth, with an average 158 lightning bolts per square kilometer striking every year. That's a lot of lightning to dodge.

Why can some people hear colors?

It's called synesthesia, and is probably best described as an involuntary combination of the senses. Hearing colors is probably the best-known example, but it can take a number of forms: Experiencing sounds as smells, matching textures with emotions, and sensing words as tastes are just three others.

Brain Games

It was first identified in the nineteenth century and it's been studied at points ever since, but synesthesia hasn't yet been explained, although there are several different theories for what causes it. There's a tendency for it to run in families, which implies that there may be some genetic link. One argument is that it results from neural connections between those areas of the brain dedicated to the senses that are usually distinct (sound having additional connections to sight, for example). Another idea is that everyone has the potential neural pathways to experience synesthesia, but that, for reasons as yet unknown, the brains of only a small percentage of people use them. However, it's clear that a lot more research will need to be done before there's a theory on which everyone can agree.

Language of the Senses

Perhaps the most interesting recent study made a connection between language and synesthesia. People who learn a second language early in life, but who aren't bilingual (that is, they don't learn to speak two languages simultaneously from infancy) are more likely to be synesthetic. This suggests that synesthesia may be some kind of mental reaction to complex learning processes, such as mastering grammar or learning music—a deep-brain equivalent to those kindergarten alphabets where, say, all the Bs are red, while the Rs are all colored blue.

Why does banging your funny bone or stubbing your toe hurt so much?

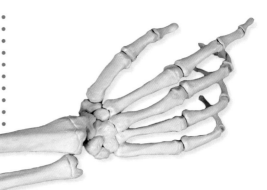

It's short, sharp, and absolutely agonizing—but a stubbed toe or a bruised elbow isn't a serious injury. So why, if you do either, does it cause such extreme pain? The short answer is that it's to do with the specific locations of nerve endings.

All in the Nerves

Our feet work hard for us, and like most body parts that interact directly with external surfaces, they're rich in nerve endings—nociceptors—that serve both to warn and to guide. Having lots of nerves means that if your foot touches something that may be painful or dangerous, it can send a warning very fast. But your toes aren't padded, and if you're walking fast, you could be putting as much as twice your body weight behind your foot's forward motion, so when you stub your toe, it creates a perfect storm of sensation: loads of nerve endings being hit very hard and fast.

Not So Humerus

The excruciating pain when you hit your funny bone is also due to nerves, or in this case, just one: the ulnar nerve. This is a major nerve that originates in the spine and ends in the tips of your ring finger and little finger. En route, it travels down your arm, passing behind your elbow. Where your upper arm meets the radius and ulna bones of the forearm, there's a single point, called the cubital channel, where the nerve has to fit between bone and skin and is comparatively unprotected. When you hit it, you get a shooting pain straight down the nerve, from elbow to fingertips, that's almost like an electric shock.

UNEXPECTED EVENTS

The human body is complicated but, by and large, predictable. It can respond unexpectedly to surprising events, from high levels of stress to lightning strikes or space travel. What have you discovered about physical reactions to the unexpected? Test yourself to find out.

Questions

1. Whose hair is said to have gone white overnight— Mary Queen of Scots, or Marie Antoinette?

2. Where would you find the natural glycoprotein Draculin?

3. Can you name two great writers who feature cases of spontaneous human combustion in their works?

4. In which country was the first-ever official verdict of death by spontaneous combustion brought in 2011?

5. Which naval figure lost his right arm in the course of a sea battle in 1797, making him one of the most famous amputees in history?

6. Why would your blood boil if you were caught out in space without a spacesuit?

7. Where would you most likely be struck by lightning?

8. Which type of stress is benefical: a) chronic or b) eustress?

9. The cubital channel runs between your ankle and your knee—true or false?

10. What do you call the condition that causes you to hear colors, smell sounds, and taste words?

Turn to page 214 for the answers.

COULD A HUMAN BRAIN BECOME FULL?

Erkenne Dich selbst

WHAT ARE DREAMS FOR?

WHY DOES EMOTIONAL "HEARTBREAK" GIVE YOU AN ACTUAL PAIN IN THE CHEST?

IS THERE SUCH A THING AS A
NATURAL SENSE OF DIRECTION?

IN YOUR HEAD

WHAT CAUSES BRAIN FREEZE?

Could a human brain become full?

It's a familiar sensation: Your brain feels stuffed full of information, and you can't quite find the specific piece of information you want, however hard you try. But could your brain really fill up . . . to the point where it wouldn't be able to take in anymore?

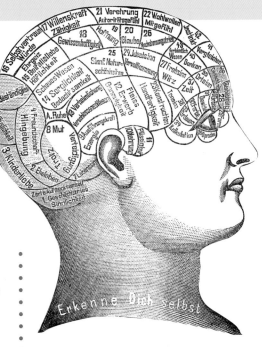

A Library Needs a Filing System

The short answer is no—because the brain isn't a petrol tank. It operates in a far more complex way than a simple "fill it up" mechanism, and it's been known for a long time that memory, key to ensuring that we think and act appropriately in different situations, is one of the most complicated systems of all, involving a number of different areas of the brain. Huge amounts of information aren't much use without an accurate method of calling up specifics (after all, the largest library in the world wouldn't be very effective without an efficient filing system).

In the human brain, one of the functions of the hippocampus, the horseshoe-shaped structure deep in the brain, is storage. But when you need a specific memory, the prefrontal cortex, which covers part of the frontal lobe of the brain, helps to filter what you need from what you don't. This means that you don't consciously have to scroll through memories every time you want a specific one.

I've Forgotten to Remember to Forget

Back in 1955 when Elvis was warbling the popular classic, no one thought that he was expressing a scientific principle rather than a romantic lyric. But it turns out that forgetting is a key part of keeping your memory efficient. With a constant "feed" of new material pouring into the brain, it's important that some of the earlier, lesser-used stuff can be put on a mental backburner to keep the memory agile.

An article in *Nature Neuroscience* in 2015 recorded an experiment that seemed to show that the prefrontal cortex strained out less relevant memories if the brain did a general search for something specific. If there were a number of similar memories, it appeared that the prefrontal cortex would ensure the right one was pushed to the fore. After the "right" memory had been found, if it was looked for again later, it attracted more brain activity than a less-used memory. Well-thumbed memories, taken out and "used" regularly, attracted the most brain activity.

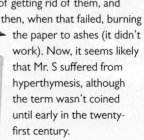

SUPER-MEMORIES

You might wish your memory was better—but what if you could remember every day of your life so far, in full, excruciating detail? People with hyperthymestic syndrome can do just that: Ask them what they had for supper on a random date 5 years ago and they'll be able to tell you, accurately and without hesitation. Alexander Luria, the Soviet psychologist and one of the founders of neuropsychology, was one of the first to record the condition when he noted that "Mr. S," a subject he studied in Moscow in the 1920s and 1930s, was desperate to discard items from his all-too-inclusive memory, first writing them down on scraps of paper in the hope of getting rid of them, and then, when that failed, burning the paper to ashes (it didn't work). Now, it seems likely that Mr. S suffered from hyperthymesis, although the term wasn't coined until early in the twenty-first century.

What causes brain freeze?

M̲ost people are familiar with the sudden sensation of "brain freeze." Also known as an "ice cream headache," this intense stabbing pain typically occurs during or after ingesting an icy-cold drink or snack. The universally experienced response is caused by the sudden change in temperature in the roof of your mouth.

Cold As Ice

The scientific name for brain freeze is sphenopalatine ganglioneuralgia, which literally translates as "nerve pain of the sphenopalatine ganglion"—the latter is the group of nerves responsible for transmitting sensations from the top of your mouth to your brain. The human mouth is surrounded by blood vessels, which contract when they are cold to prevent the further loss of heat. When you quickly eat or drink something very cold, your mouth doesn't have time to absorb the temperature very well. The roof of your mouth is most affected, because that's where your internal carotid artery (the one that feeds blood to your brain) and your anterior cerebral artery (which runs along the front of your brain, sitting on the brain tissue) meet.

Bloody Freezing

When the temperature drops, these two important blood vessels contract. Your brain sends extra blood to them to warm things up, causing them to expand rapidly. These quick changes in blood supply are sensed by the meninges—pain receptors in the outer covering of the brain. The pain signal travels through the trigeminal nerve, one of the most widely distributed nerves in the head. One theory as to why you feel pain in your forehead or behind your eyes (rather than in the roof of your mouth) is that it is referred pain—which manifests itself further along the nerve pathway. A second theory is that the altered blood-flow throughout the brain causes the pulsating headache. Brain freeze is so common and easy to

induce in subjects that scientists are using it to help them understand and create treatments for other types of headaches, like migraines.

How to Cure Brain Freeze

Vasodilation (by which blood vessels expand to increase blood flow) is thought to be a defense mechanism enacted by the body to help protect the brain from getting too cold. Brain freeze usually passes fairly quickly, but there are a few other things you can do to help your brain out. First, avoid drinking or eating really cold substances (or, if you must, sip and nibble, rather than gulping and gobbling them); second, drink some tepid water or put your tongue to the roof of your mouth to warm it up; and third, take some rapid breaths while covering your open mouth and nose with your hands—this will increase the flow of warm air to the palate.

HEADACHE HISTORY

Even though brain freeze has been discussed in medical literature since at least the 1850s, it got its first popular reference when *We Didn't Ask Utopia: A Quaker Family in Soviet Russia* was published in 1939. The author, Rebecca Timbres, was married to Harry Timbres, an American doctor who, in 1936, moved his family to Russia, where they hoped to help the fight against preventable diseases. She wrote of the cold weather, " . . . your nose and fingertips get quite numb . . . and if you don't keep rubbing your forehead, you get what we used to call 'an ice cream headache.'"

Can doing the crossword help to fend off Alzheimer's?

It's one of those pieces of received information that's probably been quoted to you many times: Keep your mind agile, do plenty of crosswords and Sudoku, and it'll reduce your chances of getting Alzheimer's disease in old age. But is it true?

Environment vs. Heredity

Only partly. This is an area that has seen several studies, and just as many inconclusive results. However, there appears to be a grain of truth in it. Around one in five people carry a gene variant—APoE4—which doubles their risk of developing Alzheimer's in

later life. And for those people, doing the crossword, Sudoku, and any other mentally stimulating activity has been shown to help delay the buildup of abnormal protein deposits (often referred to as "plaques") on the brain—and these have been strongly implicated in the development of Alzheimer's. Conclusion: Keep doing the crossword, Sudoku, and any other activity you find stimulating. It can't do you any harm.

Keep Dancing

There's no need to limit yourself to the crossword, either. A study of older people carried out by the Karolinska Institutet in Stockholm took a sample of more than 1,200 subjects aged between sixty and seventy-seven in Finland, and turned the lives of half of them upside down—in a positive way—across a 2-year period between 2009 and 2011. They were given exercise and aerobics classes and brain training with graded computer games, while their diet was overhauled to include plenty of vegetables, fish, and healthy oils. Perhaps unsurprisingly, at the end of the period, the revamped sample did substantially better than the control subjects in a range of tests, including

the research, suggested that in the absence of "management" to ensure an all-around lifestyle change, older people should take on extra or different activities one at a time. Her top recommendation for a new activity? Dancing. It's often a less intimidating option for the elderly than going to the gym, plus it can offer fun and some social interaction, as well as exercise. So if you've already taken on some mentally stimulating options, consider adding a weekly dance class, too.

HIGHER EDUCATION

There are increasing indications that a higher degree of education may also help the susceptible to dodge a bullet when it comes to Alzheimer's. Another study by the Karolinksa Institutet found a correlation between the odds for contracting Alzheimer's disease and the level of education reached by the subject. Each year of additional education past the primary stage resulted in lower odds for Alzheimer's, all the way up to completing a university degree.

an 85 percent improvement in tests of their brains' organization of thought processes, and a staggering 120 percent in the speed at which they could process information. The control group, who had been given straightforward health advice of the kind typically offered in a doctor's office, but nothing else, remained at the cognition levels of their original tests. Asked what the takeaway from the study was, Professor Kivipelto, who led

Could you fool a lie detector?

It's been a staple in movies and reality television shows for years, and regularly crops up in police procedurals, but how reliable is a polygraph or lie detector test—and could you learn to fool one? Veterans of the procedure claim that you don't have to be a sociopath to cheat successfully.

Ordeal by Blood Pressure

The first machine, a modified blood-pressure cuff, was created in 1917 by William Moulton Marston, a psychologist, lawyer, and inventor (and incidentally, under his pen name, Charles Moulton, the creator of Wonder Woman). It was followed up 3 years later by the first named "polygraph" (from the Greek, "many languages"), the invention of John Augustus Larson, a California policeman, which not only measured physiological changes in subjects but also plotted them on paper.

How Do They Work?

The principle behind modern lie detectors is that, having set the level at which the subject is physically comfortable with a range of test questions, the examiners have a picture of his or her "norm," after which the answers to leading questions can be assessed against a baseline of the subject's regular breathing rate, perspiration levels, pulse, and blood pressure. Modern polygraphs include "skin conductivity," the phenomenon by which the skin becomes a more efficient conductor of electricity when its owner is subjected to stress. Some versions are trialing MRI results, too.

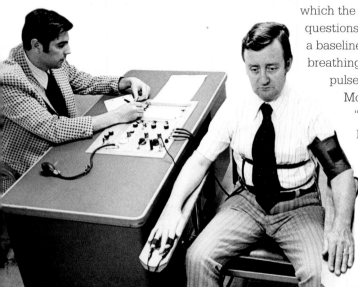

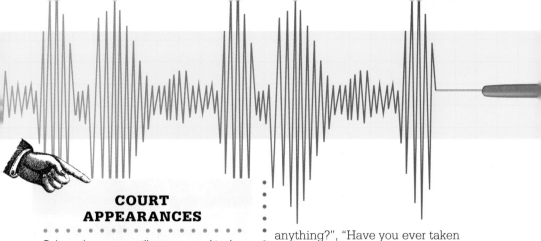

COURT APPEARANCES

Polygraph tests are still not accepted in the vast majority of criminal courts. They're used in other applications, though—in interviews for candidates for security jobs, for example. In some cases, convicted criminals have taken lie detector tests in attempts to prove their innocence. Nevertheless, currently there seems no prospect of lie detector tests becoming widely accepted as evidence in court.

Setting the Bar

Polygraph tests are conducted by skilled operators with a good deal of training in interpreting the results (which are rarely as cut-and-dried as you might imagine). Sessions are usually preceded by a general chat between subject and operator. When the test starts, the initial questions are innocuous. They're usually followed by some "control" questions, in the course of which it's assumed that the subject will tell at least one white lie. They will include a wide range of topics, some leading, such as "Have you ever stolen anything?", "Have you ever taken anything from a shop without paying?", and so on. Oddly, you might think, most people will lie at some point— that is, they'll answer "No" to one of these questions, despite the fact that, as the operator knows, almost everyone has stolen something, however small, at some point in their lives. Despite the "No," that's the point at which physical signs will indicate on the polygraph that you are lying, after which the tester will start to incorporate more leading questions into the mix.

So How Can I Cheat?

Skeptics claim that anyone can trick a lie detector. The thinking goes that you can cheat—by doing a complicated math problem in your head, for example, or, as in the movie *Ocean's Eleven*, by pressing your foot down on a thumbtack in your shoe to set your baseline at higher levels than are natural to you. Subsequently, when you lie, it won't create the same peaks on the polygraph.

Is there such a thing as a natural sense of direction?

We all know someone who seems to have an innate sense of where they are, even on unfamiliar streets. But is this something that comes naturally to them, or is it a practiced skill, or a bit of both?

In the Cells

Along with other mammals, every human has a built-in GPS (global positioning system)—neurons whose job it is to help you navigate your way around. In the hippocampus, deep in your brain, you have place cells. Among other functions, the hippocampus is responsible for your memory, and place cells are believed to give you a sense, specifically, of where you are in your environment, and to help you make a mental map of your surroundings. Next-door to the hippocampus is an area called the entorhinal cortex, which contains two other types of cells whose jobs, together with that of the place cells, are believed to add up to a sense of direction. The two types are grid cells, which help you "map" as you move between locations, and head-direction cells, which—as you'd expect—give you a sense of which way you're facing. The roles of all three are complicated and as yet far from fully understood, but it's thought that grid cells and head-direction cells work in tandem.

Knowing Your Way Around

So far, so good: The three types of cells give you a basic navigational sense. But why are some people so much more effective at finding their way around than others? It may have something to do with the strength of the signals between cells, particularly those in the entorhinal cortex. We all have these signals, but in some people they're stronger than others. The hippocampus and entorhinal cortex are two of the areas most frequently undermined in cases of Alzheimer's—and it's been suggested that this may be one of the reasons that losing a sense of direction is often one of the early signs of the onset of the disease.

Hone That Skill

Even the most phenomenal "natural" senses may dim if they're never used. Recent studies into the scenting capacity of dogs have shown that lack of practice over several generations (such as in dogs who have all their needs met, and who have constant interactions with humans, for whom smell isn't an important part of games or activities) may dull a sense that, in dogs, is inherently strong. Some experiments with GPS have shown that something similar may happen with humans and a sense of direction. A constant reliance on GPS makes users less aware of the environment around them, and they don't "micro-navigate" (that is, use specific landmarks to work out the next step on a journey) because the GPS renders it unnecessary. So if you're worried that your sense of direction isn't all it should be, try navigating some journeys by using a map, rather than leaving it all to the dulcet tones of your satnav. London taxi drivers are famous for holding mental maps of huge areas of London, worked and honed by doing "the Knowledge," the qualifying exam for a taxi license. It's probably no coincidence that in a study made by University College London their hippocampuses were found to be bigger than average.

Are some people preprogrammed to be addicts?

A group of people enjoy a regular drink or experiment with drugs, but only one of them becomes an alcoholic or a drug addict. Why? In the past, addictions were put down to moral weakness; now, while the mental circumstances around addictions aren't completely understood, it's believed that an addictive tendency is made up of a complex combination of genetics, environment, and psychology.

A Perfect Storm

No single factor decides addiction. It's true that there seems to be a genetic component. One large Australian study discovered that, while problem gamblers were more likely to have a certain gene, some gambling addicts didn't have it, and some people who had it hadn't developed a problem with gambling. However, it's not fixed—it's a predisposition, not a prediction. Environment also plays a part. The children of addicts —for whom, if they are brought up by their parents, addiction may be part of both their environment and heredity—are generally more likely to become addicts themselves, but, again, not *all* the children of addicts become addicted. And there's the psychological and chemical aspect. The part of the brain most strongly implicated in addiction is called the limbic system. It's a complicated group of structures deep in the brain, and is often called the brain's "reward system." It has numerous functions, but one theory of addiction is that some people have a more sensitive limbic system than others, which may cause a predisposition to become hooked on the chemical results of some substances. If all three switches—environmental, genetic, and psychological—are flicked, then the likely result is addiction.

Changing the Brain

The needing-a-"fix" quality of addiction comes from the neurotransmitters that a dose of something—drugs, alcohol, even gambling—causes to be released. The neurotransmitters attach themselves to receptor cells, but while they naturally only remain there for a brief time before breaking down, addictive substances interfere with

WHISKY

DOPAMINE: THE CELEBRITY NEURO-TRANSMITTER

The main neurotransmitter in the brain's "reward" pathway is dopamine, although it's not the only one. Over a hundred different neurotransmitters have been identified, each with a different job to do. In a 2013 article in *The Guardian* newspaper, Vaughan Bell, a neuropsychologist at University College London, ironically dubbed dopamine "the Kim Kardashian of neurotransmitters," querying its party-animal reputation, and pointing out that it plays many roles according to what part of the brain it's in. The article was making fun of the increasing tendency for pop-science writing to dump almost any addictive tendency at dopamine's door: Claim that a substance elevates dopamine levels, and it's immediately declared dangerously addictive. And while that's true to a point, it's certainly not the whole picture.

the way the cells deal with dopamine in various ways. At first addiction experiences create a pleasant sensation; over time, with repeated use of the stimulant, the brain adjusts, and neurotransmitters are either released in a smaller quantity or the responses of cells sensitive to them become dulled. This sets up a vicious cycle—a dose that previously saturated the system will soon cease to be enough to access the same sensations—which is what makes addiction particularly hard to beat.

Does referred pain have a function?

You've probably heard that a heart attack can be preceded by a bad pain, not in the chest, but instead in the left shoulder and arm, or between the shoulder blades. This is known as referred pain, and is the result of a mix-up of signals sent by the brain—but the problem originates in the heart. So has the brain simply made a mistake?

How it Happens

Referred pain seems to happen when pain signals are sent from one site on the body into the spinal cord, and somehow stimulate nerves that aren't directly affected by the source of the pain, causing the sensation of pain to be felt somewhere else. It's a phenomenon that isn't yet fully understood, but it does seem to stem from a muddle in the brain. The pain itself should serve the usual function of pain—to alert the body's owner to a problem that needs attention—but in the absence of proper location co-ordinates, referred pain can be a problem for health professionals, sending them off on a hunt around the body to find what's really wrong.

THE ADVANTAGES OF PAIN

Difficult though referred pain can make things for the sufferer, it's a very small problem indeed compared to how life is for those few people who don't feel pain. This condition, CIP, or congenital insensitivity to pain, was first identified in 1954, and is fantastically rare—just a few hundred people in the world are believed to have it at any one time. It occasionally crops up as an attribute in horror films: The almost superhuman figure who, because he feels no pain, is an invincible opponent. But the real-life situation for those with CIP is very different. The function of pain is to warn you not to put yourself in harm's way; with no pain to act as a deterrent, people with CIP constantly behave in highly risky ways (in fact many die young, in accidents, simply because they couldn't learn caution in the absence of any warning signals). Far from being a superhuman power, the freedom from pain is, in reality, a spectaculary risky attribute.

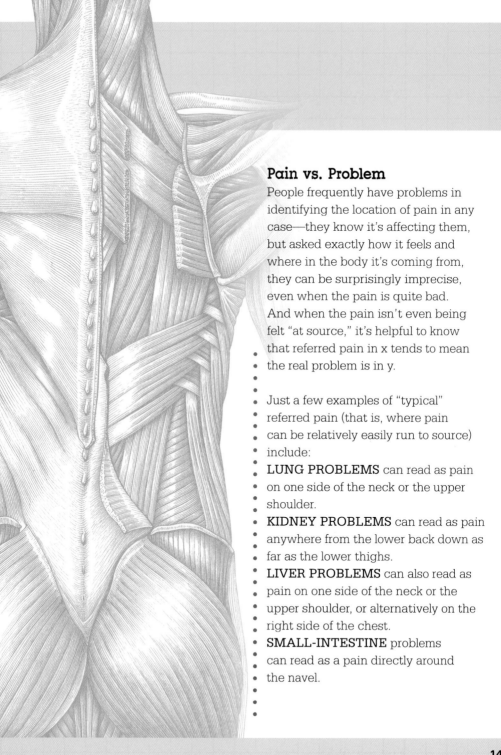

Pain vs. Problem

People frequently have problems in identifying the location of pain in any case—they know it's affecting them, but asked exactly how it feels and where in the body it's coming from, they can be surprisingly imprecise, even when the pain is quite bad. And when the pain isn't even being felt "at source," it's helpful to know that referred pain in x tends to mean the real problem is in y.

Just a few examples of "typical" referred pain (that is, where pain can be relatively easily run to source) include:

LUNG PROBLEMS can read as pain on one side of the neck or the upper shoulder.

KIDNEY PROBLEMS can read as pain anywhere from the lower back down as far as the lower thighs.

LIVER PROBLEMS can also read as pain on one side of the neck or the upper shoulder, or alternatively on the right side of the chest.

SMALL-INTESTINE problems can read as a pain directly around the navel.

What are dreams for?

It's hard to say definitively what dreams are for. They've been the subject of both amateur and professional interest for centuries, yet despite being so widely studied, they remain obstinately open to a vast number of interpretations.

Everyone Dreams

Even when you don't remember them on waking (it's believed that the dreamer won't have any recollection of 90 percent of their lifetime's dreams), it's estimated that you probably have somewhere between three and six episodes of dreaming every night. How do we know? Because many, many studies have established that between 20 and 25 percent of your sleeping time is spent in REM (rapid eye movement) sleep, and REM sleep indicates that you're dreaming.

Theories of Dreaming

Sigmund Freud, the father of psychoanalysis, was the first modern great to put forward a theory of dreams. In *The Interpretation of Dreams*, published in 1899, he held that they were fundamentally expressions of wish fulfilment: everyone's chance to explore the thoughts that they might

not dare express, or even consciously feel, when awake.

In the 120-odd years since his magnum opus was published, Freud's theories have been debated, dismissed, and reinstated in a recurring cycle. But even though some people disagree with him, and many have over the years, he set the scene for debate.

THE KNACK OF NIGHTMARE AVOIDANCE

The old myth that eating cheese before bedtime will give you bad nightmares is just that—a myth. But there's some research that seems to suggest that going to sleep early may reduce the likelihood of bad dreams. In 2010, two studies, one from Turkey and the other from Canada, found that people who went to bed very late reported a higher incidence of nightmares. It's not clear why this should be, although one theory blames a natural rise in your cortisol levels in the early morning. On an ordinary schedule, this would be just before you wake up, but if you were very late to bed, the rise may happen while you're still in REM sleep—and could be a prompt for unusually vivid, bizarre, or just plain horrible dreams.

Tried Favorites

There are literally dozens of extant theories on the purpose of dreams. Just three current favorites:

There's the snappily named activation-synthesis hypothesis. This holds that dreams don't really have an inherent meaning: Instead, they're a random selection from our thoughts and feelings that are constructed by electrical impulses in the brain. When we wake up, proponents of this theory hold, our conscious minds try to turn this jumble into "stories" in an attempt to make sense of them.

Another theory is that dreams are a by-product of the brain processing information. The idea goes that as we sleep, the brain works to make sense of, and store, all the information that it has acquired during the preceding day, and dreams are either the waste material from this processing or possibly even an as-yet-little-understood stage in it.

There's also the threat-simulation theory. This explains dreams as a sort of simulation process. It suggests that threatening dreams give humans a mental rehearsal for difficult situations that might arise in real life—helping to ensure that they make the most pragmatic decisions when such situations arise when they're awake. This explanation handily covers the fact that animals other than humans dream, too.

Why does emotional "heartbreak" give you an actual pain in the chest?

We know—or think we know—that pain is really the business of the brain. So how is it that when you feel deeply emotionally upset, it's often experienced as a literal ache in the chest, as though the pain were actually located in your heart?

Brain Pain

There are numerous terms for it, and we use them all the time. We refer to "heartache"; we say we feel "heavyhearted"; and when things are really bad, we're "heartbroken." Pain, although it can be felt all over the body, originates in the brain, in an area called the anterior cingulate cortex, and the same brain area deals with both physical and emotional pain, without apparently distinguishing between the two. We know that emotional distress can hike your heart rate, cause your muscles to tense, and make you feel queasy in the stomach.

Chemicals for Heartbreak

Chemicals play a part. The "feel-good" chemicals of dopamine and oxytocin that supply all-around feelings of emotional well-being are replaced when you're unhappy and stressed. Cortisol and adrenalin, which move into your system instead, have much less-benign effects. They can cause your muscles to tense up—the fight-or-flight response—resulting in that uncomfortable tight feeling in your chest.

Nervous Stress

Exaggerated stress in the anterior cingulate cortex may also provoke it to prompt extra activity in the vagus nerve, a major neurological pathway that runs from the brain stem down through your neck, chest, and abdomen. If it's overstressed, the vagus nerve can cause both physical pain and nausea. And this may be what's happening when you feel—literally—that you're heartsick.

QUESTION 67

If you really have a headache, is it better to have sex, or to avoid it?

It's the most tired excuse in the world if you don't want an intimate encounter, but if you really have a headache, will having sex help to alleviate the pain, or could it actually make it worse?

Better Than Painkillers

For some kinds of headache, sex is beneficial. It can help to relieve the pain of migraine or cluster headaches (to the point at which some sufferers actually use lovemaking as a form of pain relief). A study of a large group of headache sufferers, made at the University of Munster in Germany in 2013, showed that over half of the migraine-prone subjects found that having sex relieved the pain of a migraine—in one in five cases, it got rid of it altogether.

Coming and Going

How does having sex dissipate a migraine? It may be the orgasm, rather than the whole sexual act, that has the good effect, causing a rush of endorphins through the central nervous system, blunting the pain by blocking its messages from accessing the brain.

WHEN SEX *CAUSES* HEADACHES

Although the news sounds good, there's a downside, too. For some people the sex/headache equation seems to work the other way around: They suffer from acute headaches directly after having sex. Experts attribute this to two possible causes—first, sufferers may experience pressure on their back or neck as a result of the position they hold during sex, and this can be a cause of headaches; second, there's the straightforwardly named "coital" headache, caused by the dilation of the blood vessels that accompanies orgasm. So, can sex cure a headache? Yes, but it seems that it can cause one, too.

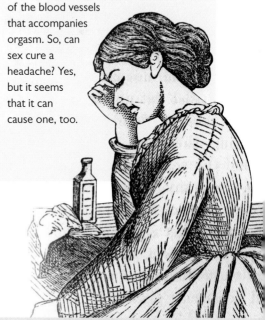

How many watts does your brain run on?

Your brain uses more energy than any other body organ—anything up to 20 percent of the total amount your body generates. So what does that mean in wattage terms? Could your brain energy power a light bulb, for example?

Sharp Brain, Dim Bulb

The answer is pleasingly precise: 12.6 watts. That figure was published in an article by Ferris Jabr in *Scientific American* in 2012. He made the calculation by converting the calories used by an average person's daily resting metabolic rate (around 1,300 kilocalories) into joules, then from joules to watts, before finally dividing the total (63 watts) by five (20 percent). That's not very much in commercial energy terms—think of the dim light supplied by, say, a 25-watt bulb. Yet that's double the wattage of the human brain.

What's All That Energy For?

A 2008 study at the University of Minnesota Medical School found that two-thirds of the brain's energy requirements are used in communication—ensuring neurons send the signals necessary to meet all the jobs the brain does; the remaining one-third is dedicated to maintenance, more of which takes place during the body's downtime.

JOBS YOU NEVER THOUGHT OF

The mini-wattage of the brain becomes more intriguing when you consider that it not only looks after the heavy labor of thinking and body management, but also handles dozens of chores you've probably never considered. Why, for example, doesn't the world flicker or go dark every time you blink? Because your brain seamlessly fills in the mini-gaps, recording what you're seeing as the eyelid goes down, then linking it to the picture your eye sees as it goes up again.

IN YOUR HEAD

It only weighs around 3 pounds, yet the human brain works ceaselessly as the control center for the rest of the body. Take this quiz to find out whether you're a dim bulb or a brain box.

Questions

1. If you suffer from hyperthymestic syndrome, are you a) especially forgetful, or b) in possession of a super-memory?

2. William Moulton Marston invented the first primitive lie-detector. What else is he also known for?

3. Which neurotransmitter was hailed as the "Kim Kardashian of neurotransmitters" in a newspaper article?

4. Which condition do the initials CIP denote?

5. Why doesn't the world go dark when you blink?

6. Sphenopalatine ganglioneuralgia is the scientific name for a migraine—true or false?

7. Are highly educated people less likely to get Alzheimer's Disease?

8. If your place cells, grid cells, and head-direction cells are working well together, what are you able to do?

9. If you have a pain in the neck, it may be a symptom of a problem with your liver—true or false?

10. What does it mean when a sleeper's eyeballs move rapidly under their closed eyelids?

Turn to page 215 for the answers.

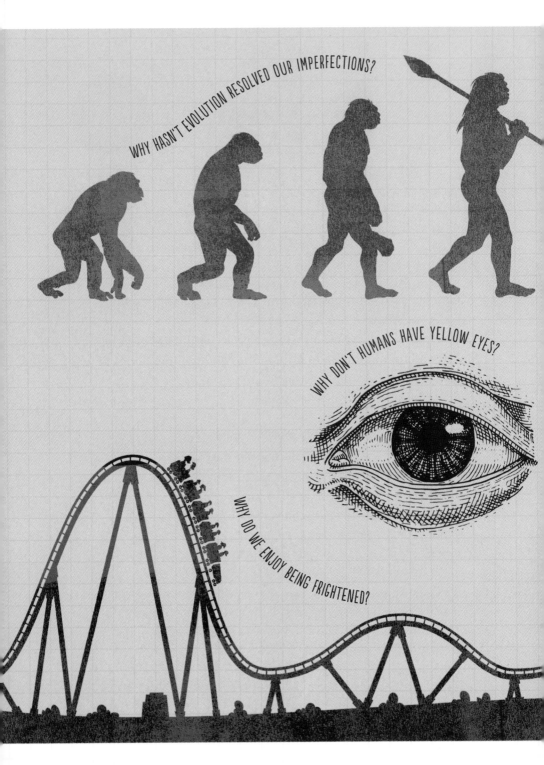

WHY HASN'T EVOLUTION RESOLVED OUR IMPERFECTIONS?

WHY DON'T HUMANS HAVE YELLOW EYES?

WHY DO WE ENJOY BEING FRIGHTENED?

HOW MANY SMELLS CAN YOUR NOSE IDENTIFY?

CAUSE AND EFFECT

WHY CAN'T YOU TICKLE YOURSELF?

How many smells can your nose identify?

You may already know what range of sounds your hearing covers, and how good your sight is, but it's doubtful that you'd be able to say with any degree of certainty how many different scents you're able to pinpoint.

A Trillion Smells

Up until a few years ago, the scientific view of a human's smelling ability rested at a rather vague "around 10,000" scents. That number was arrived at in the course of an experiment conducted back in 1927, in which four basic smells (if you're curious, they were fragrant, acidic, caproic—a technical term for a musky or sweaty smell—and burned) were tested in different combinations at ten different concentrations each. The result was broadly accepted for more than 80 years. But an experiment at Rockefeller University's Laboratory of Neurogenetics and Behavior, New York, in 2014 was to change that presumption. Its findings, immediately trumpeted across the scientific media, were that the average human can identify a trillion different smells (if that number's too large to compute, it's a million-times-a-million). It turned the official view of human scenting ability on its head.

Testing It

The test worked by isolating 128 different scent molecules and presenting mixtures of them—in combinations of ten, twenty, or thirty different molecules—to the study's subjects, three "smells" at a time. In each case, two of the scents would be identical, and a third made from a

different combination. Even in cases where all three mixtures had over half the same components, the people taking part in the test could still pull out which smell of the three was different. The figure of a trillion smells was worked out by multiplying all the possible combinations.

Unfortunately, a year or two after the experiment had taken place and the results were published, other scientists queried the mathematical calculations behind it, calling the results into question. But even if it's eventually found that a trillion was a flawed estimate, the experiment established how much richer our sense of smell is than previously thought.

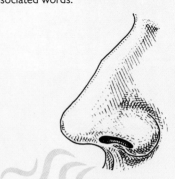

What's the physical point of crying?

It all depends on which tears we're talking about: Humans produce three types, each for a different reason—basal tears, reflex tears, and emotional tears. The first two have quite distinct physical functions, the third type is rather more complicated.

What's the Recipe?

All your tears are produced by the lacrimal glands, which are located just above each eye. They travel through ducts to the inner corner of each eye, where they're released. They're made from more than just salty water—every tear contains vitamins and minerals, oil, mucus, and lysozymes, which are a form of natural antibiotic. The mucus helps the tear to adhere to the eyeball, the lysozymes are there to fight off any potential eye infections, and the oil stops the tears from evaporating and allowing your eyes to dry out.

Tears at Work

You produce basal tears all the time, in a quantity that is just enough to keep your eye lubricated. Reflex tears are different: They're produced in reaction to an irritation—for example, if a speck of dirt in your eye needs to be removed. Unlike basal tears, reflex tears are produced in copious amounts to stand the best chance of washing the intruding substance out.

Emotional tears are the tears you produce when you're stressed or upset (or, perhaps more rarely, overcome with joy). A long-disproved but charming theory about the origins of emotional tears was widely held in the eighteenth century. The warmth of your feelings made your heart overheat, it was believed, and water vapor was then generated to cool the organ down.

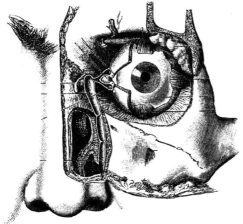

The vapor rose up through the body, before condensing back to liquid water and emerging as drops through the eyes. It's such a neat and appealing idea that it seems a pity that it's not true. It is now known that they are sparked when the limbic system in your brain, which manages your feelings, activates a neurotransmitter called acetylcholine—and this prompts your lacrimal glands to produce tears. It's harder to say why, although there are various theories. One is that emotional tears evolved to act as a social distress call, inciting those around you to offer support when you're feeling upset. Another is that, for you, the crier, crying is cathartic. It deals with difficult emotions, clears

WHAT ABOUT ONIONS?

If the reason for reflex tears is to protect the eye, why, you might ask, do onions make you cry? Actually, that's to protect the onion. Alliums produce their sharp scent (a chemical called propanethial-S-oxide) in an attempt to avoid being eaten. If an animal crunches into a raw bulb only to find that its eyes are assailed by a painful stinging sensation, then it's likely to give up on this aggressive food choice. Cooking the onions gets rid of both the chemical and the effect—humans: 1, alliums: 0.

the decks psychologically, and leaves you more ready to cope. The catharsis idea is supported by the fact that your emotional tears have a different chemical composition from the others. They contain more manganese (excessive manganese levels have been linked to depression) and stress hormones than basal or reflex tears. At least in part, they may be your body's way of getting rid of emotionally aggravating chemicals.

If all your cells are renewed every 7 years, why do you age?

Well, for a start, the almost universally quoted idea that all the cells in your body are renewed every 7 years isn't true. Where did it come from? Nobody knows, but both this idea and a common variation (that your body renews itself completely every 10 years) still make regular appearances.

A Ripe Old Age

Cells do have a natural life span, but it's very variable. Some are expected to last you a lifetime—you are born with all the brain cells you will ever have, for example, and they aren't replaced when they die off. At the other end of the scale, the cells lining your gut live for just 3 or 4 days. And there are many variations in between—cells in the alveoli of your lungs live for about a week, red blood cells for around 4 months, while fat cells may live as long as 8 years. It's a gruesome but fascinating truth that some of your cells will even outlive you: When your heart stops beating and you stop breathing, living cells somewhere in your body may linger on for another few hours or even a day or two.

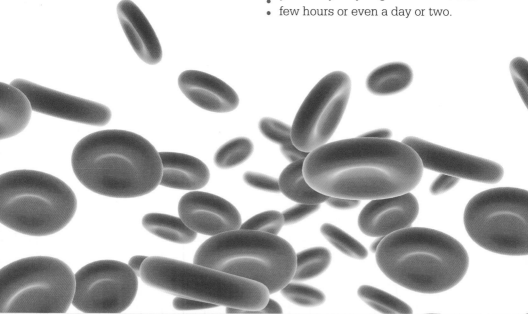

HOW LONG DOES A TATTOO LAST?

It may blur a bit, but unless you have your tattoo removed, it will be with you for life. This is because the tattoo needle is pressed deeply into the skin, going below the surface layer of skin (the epidermis) and into the dermis, the layer underneath. Cells on the epidermis are abraded and renewed all the time, but those in the dermis aren't replaced so quickly, and the ink fragments that make up your tattoo, while tiny by your standards, are still too large to be "tidied away" by the white blood cells whose job it is. If you want a tattoo effectively removed, the only way to do it is by laser. The laser beam blows the ink fragments into minute pieces—and they become small enough to be removed by the tidying cells on their regular rubbish-collecting patrols.

Ending Up

Cells renew by making copies of themselves—they split, creating an identical new cell, a process called mitosis. From infancy to adulthood, as long as you're growing, your cells will be multiplying; once you're fully grown, though, your cells will only split to replace damaged or dead cells. But cells do age. Every cell contains DNA, and every DNA strand is topped off by an essential "cap" called a telomere. Each time a cell replicates itself, the telomere gets a little shorter, and as the cell becomes increasingly elderly, the telomere eventually gets too short for the cell to be able to divide—and at this point the cell will die. Since we're made of cells, and cells themselves get old, we get older with them.

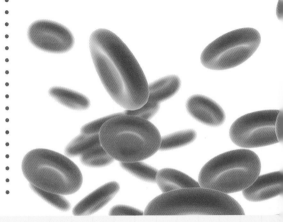

Why can't you tickle yourself?

Small children can be driven into ecstasies of hysterical pleasure by being tickled; as we age, we tend to stop finding the sensation so enjoyable (maybe we become more fearful about losing control or dignity?)—and some of us actively hate it. However, one thing every child knows is that it's hard-to-impossible to tickle yourself. Why?

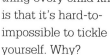

A Question of Expectation

Although it seems like a simple question, this raises some quite complicated issues about how you experience yourself as a separate entity from others. The idea that you know that you're you is so entrenched that it appears obvious to us, but it's impossible, so far, to give any form of artificial intelligence the same sense of self, so it's clearly more complex than it might appear. An experiment conducted at the Institute of Cognitive Neuroscience at University College London looked at the different brain responses elicited when people were tickled by others, and when they tried to tickle themselves.

Who's Touching You?

The experiment found that the brain creates a very clear map of your movements that simultaneously sends signals to the somatosensory cortex, where touch sensations are experienced, to "warn" it that the movements are your own and not anybody else's, and to the anterior cingulate cortex, which processes pleasurable sensations. The somatosensory cortex message ensures that you won't react when your own

hand brushes your knee in the way that you would if it were someone else's hand. You know where your own touch will land and you have an expectation of it, whereas you don't have the same expectation of anyone else's movements.

The same study tried to set up a situation in which its subjects could fool their internal system and learn to tickle themselves. One experiment required them to move a lever that would cause their hand to be stroked—but at minutely different intervals. It proved that they could tickle themselves once an unpredictable delay was introduced, because they could no longer precisely anticipate when the ticklish trigger was going to arrive.

TWO KINDS OF LAUGHTER

Scientists have identified two categories of tickling—knismesis and gargalesis. Knismesis is the light "ticklish" sensation you get when a feather brushes against your hand—it's unlikely to make you laugh. Gargalesis, on the other hand, is what might be called "heavy" tickling: The kind that drives you into fits of semi-reluctant laughter. Not only are there two types of tickling; there are also two kinds of laughter. When heavy-handed tickling does make you laugh, it won't be in the same way you might laugh if you heard a funny anecdote. A study at the University of Tübingen, Germany, in 2013 found that while both kinds of laughter are felt in the Rolandic operculum (an area of your brain that controls emotional responses and the way your face moves in reaction to them), only the tickling-induced laughter also prompts a response from the hypothalamus, the region that can trigger your adrenalin-loaded flight-or-fight response and alert you to danger.

Why hasn't evolution resolved our imperfections?

If it is evolution that has brought us from apes that learned to walk upright all the way to present-day *Homo sapiens*, and has also ensured that humans have mastered the immediate challenges of their environment, why do we still have wisdom teeth? And a coccyx? Why aren't we perfect yet?

What Evolution Does

The idea that the process of evolution is a constant quest to produce some kind of physically superior superhuman organism is misguided: That simply isn't how it works. Evolution, in the Darwinian model of the survival of the fittest, is a process by which an organism can develop to fit its environment. It's not a higher power that looks at an objective "best effort," it's more like a very, very slow conveyor belt in the course of which features that aren't fit for purpose, or that impede a species from achieving maturity and breeding, are very gradually removed.

If we were still depending on our teeth to rip up raw carcasses, our wisdom teeth—originally heavy-duty "third molars," which were useful in chewing large pieces of meat in the absence of knives to cut them up—might not have become redundant, but human evolution took a different path: We learned to live in communities, we learned to farm, and we invented cooking and knives. You don't need those extra teeth any more, but they're not a serious impediment to your

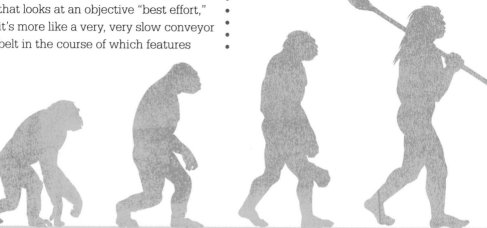

LIVING FOSSILS: FAILURE OR SUCCESS?

Maybe you've heard the astounding story of the coelacanth, the fish presumed extinct for 66 million years, until it was rediscovered off the coast of South Africa in 1939? Or perhaps you've watched YouTube movies of horseshoe crabs, primeval-looking creatures that haven't changed much from their fossil selves—which lived a staggering 148 million years ago? For years, these creatures were lumbered with Charles Darwin's "living fossil" label, with scientists concentrated on life-forms that had performed developmental somersaults along the evolutionary path, ending up looking very different from their ancestors. But does the similarity of living fossils to their distant forebears make them failures? After all, they're still living and breeding in their long-term habitats, so they could instead be striking examples of evolutionary success. A paleontologists' study of these survivors, published in 2014, coined a new term for them— stabilomorphs—which the authors felt better reflected their unique evolutionary status.

survival. Nor is a coccyx, that vestigial tail. No one is going to die because they possess either wisdom teeth or a coccyx.

Is Evolution Over?

Some scientists believe that for humans, evolution has effectively ended: They argue that, in the Western world at least, the rate of survival is now so high that there's no possibility of physical shortcomings being weeded out—"the survival of the fittest," they believe, doesn't hold when the bar is set so low. Others think that evolution will still be tested by future world-changing events— they point out that, in, evolutionary terms, what we call "the developed world" has lasted the equivalent of just a few seconds.

Why do we enjoy being frightened?

You wouldn't enjoy a real-life encounter with Freddy Krueger, so why do you get such a kick out of coming across him in a movie? In fact, why do slasher movies, dangerous sports, and roller coasters all have so many fans? And what is it that distinguishes "real" fear from the artificial sort that we seek out for the buzz it gives us?

Safe Limits

The answer lies in a mixture of chemicals and context. Context is the simpler aspect, so let's deal with it first. Enjoying a fake-dangerous experience depends on knowing that, in reality, you're safe. That's why most people can enjoy a frightening movie, at least to some extent, while far fewer will be enthused by a roller-coaster ride, and perhaps even fewer will sign up for a bungee or parachute jump. The movie is evidently not real, the other two offer a higher degree of genuine risk.

Chain Reaction

The workings of your brain do a lot to uncover the difference between real and "fake" fear. Your reaction to a real threat is the fight-or-flight response (see box at right). This is designed to make your body behave as effectively as possible when you're in danger, and it's triggered by the amygdala, the brain's emotional center. But the hippocampus, nearby, is able to gauge the degree of actual threat by putting it in context, and, if necessary, to dial back the amygdala's response and to let you know that, despite some signals, you're actually okay.

A Scale of Fear

People's limits of tolerance for what might be called "fun fear" are widely varied. The chemicals released in your brain in reaction to fear are in many cases the same as those released when you're in a state of positive excitement, and while some subjects quickly reach saturation (that is, they don't want any more of an adrenalin rush), there are others—for whom the term "adrenaline junkies" was coined—for whom the chemical rush can be addictive. It is believed that thrill seekers release greater amounts of dopamine, the "reward" neurotransmitter, from the hypothalamus, than their more cautious fellows, who don't find that risk-taking gives them the same chemical high and who therefore aren't tempted to raise the stakes in order to get a bigger hit.

FIGHT OR FLIGHT

Your reaction to a stressful event starts when your amygdala sends a signal to the hypothalamus, which forwards an alert to the adrenal glands. These respond by pushing adrenaline into your bloodstream. Once in circulation, this makes your heart beat faster and your breathing gets faster as your lungs' airways dilate to take in the maximum oxygen. Adrenalin also prompts the release of additional glucose and fat stores at multiple sites in the body into the bloodstream to give you extra energy. These responses happen instantaneously and will help you to run away or to fight, depending on the nature of the threat. If the threat persists, the hypothalamus activates a second signal to the pituitary gland, which send an additional hormone, ACTH, to the adrenal glands, prompting them to produce cortisol. This maintains the "high alert" response started by adrenaline. If the danger reduces, the cortisol levels in the body will fall and it will gradually return to normal.

Why do older people snore more?

Maybe someone has never snored, but now that they're getting older, they've started. Or they were a moderate snorer, but in old age the noise has become both irregular and deafening. But what causes it, why does it get worse with age, and why is it so difficult to fix?

How Snoring Starts

Snoring is a very common problem—a study in 1993 found that, of a large mixed sample of subjects, 44 percent of the men and 23 percent of the women tested were regular snorers. It is the result of muscles slackening while you're asleep. The muscles that line the airway and keep it open while you're awake relax as you drop off, narrowing or even partially obstructing the tube and making the airflow irregular. The air that usually flows smoothly and quietly down your throat and in and out of your lungs is disrupted by the now-variable width of the passage it passes down, and this creates irregular puffs and blasts of air, which—when they hit the slackened muscles in the throat—result in the uneven-bellows effect of snoring. Heavier people will often snore more loudly than slimmer ones, too, because fat deposits stored in the walls of the airway vibrate in a similar way to loose muscles. And snoring may be aggravated as you age, as muscles will get floppier the older you get.

DIDGERIDOO THERAPY

What's the strangest solution to snoring ever proposed? Learning to play the didgeridoo can help. A study carried out in 2006 in Switzerland gave patients at risk of sleep apnea regular didgeridoo lessons across a couple of months. Subjects were required to practice for at least 20 minutes a day, five times a week. And the results were positive—not only was daytime drowsiness (a common symptom of sleep apnea) reduced, but subjects' partners reported greatly decreased disturbance overnight, too. Why should it work? To play the didgeridoo, the musician must be able to practice circular breathing—that is, to inhale through the nose while controlling airflow into the instrument from the mouth—and this tightens and tones the muscles of the airways, making it less likely that they'll "flop" overnight.

How to Stop It

There are numerous suggestions to help stop snoring, some more eccentric than others. Sleeping on your side, rather than your back, is less likely to obstruct the airway; one uncomfortable-sounding suggestion is to duct-tape a tennis ball to your back so that you don't turn onto your back while you're asleep.

Fatal Snoring

Snoring is irritating, but it can be dangerous, too. Bad snoring may be a symptom of sleep apnea, a condition that can cause sufferers to stop breathing altogether as they sleep. This may happen very frequently—in really bad cases, every few minutes—and as the body becomes aware that it's running short of oxygen, it startles the sleeper awake. This can cause disruption, sometimes disastrous disruption, to sleep patterns. It's even been implicated in a higher incidence of heart attacks and strokes, so if diagnosed it should be taken seriously.

Why don't humans have yellow eyes?

Brown, blue, gray, occasionally green or hazel—why is the human eye palette so limited? Why don't you ever see someone with emerald eyes, like a cat? And why doesn't anyone have golden or bright red eyes?

In the Genes

The color of your eye is decided genetically, and created by the pigmentation of the iris, the ring of muscle around the pupil. Humans are unusual in having a range of possible iris colors in the first place. Variable eye color in a single species is extraordinarily rare in wild animals, although it does exist in some domesticated species such as dogs— no one is quite sure why, but it is known that in evolutionary terms these variations are comparatively recent.

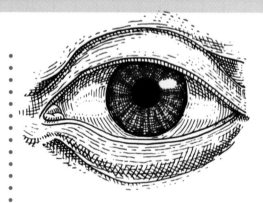

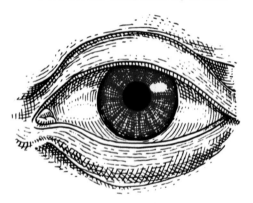

Melanins and Eye Color

Your eye color comes from melanins, pigments that also fix your skin and hair color. There are three variants: black and brown eumelanin, which cover the darker spectrum, and pheomelanin, also called lipochrome, which is on a range of red–yellow. Eye color is decided by how much of which melanin you have in which part of the eye. All eyes have eumelanin at the back of the iris, but it's the type and amount at the front of the iris that decides which color your eyes are perceived to be—and at what depth. Brown eyes, for example, have a high amount of eumelanin at the front of the iris, while hazel ones have a balance of eumelanin and pheomelanin. Blue eyes have very low melanin levels; the "blue" is actually blue light reflected out from them, rather than a blue coloration in the iris itself.

BABY BLUE EYES

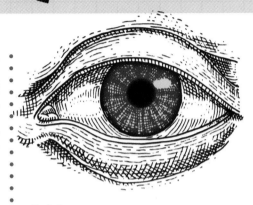

Even though only around 20 percent of Caucasian adults are blue-eyed, most Caucasian babies are born with blue eyes, which, if they're not destined to have blue eyes as grown-ups, gradually change over the months after birth. And while many darker-skinned babies are brown-eyed from the start, a few also have blue eyes that, again, turn brown while they're still infants. Why so blue?

It's because, whatever race you are, you aren't born with your full quota of melanins. Babies are born with a relatively small quantity of melanin, a little of which is in their irises—enough to make them appear pale blue—but melanin will go on increasing in their systems after birth. Even babies who have brown eyes from the start, indicating that their base quantity of eumelanin is higher, won't reach the melanin levels they'll have as adults until they're around 2 years old.

Goldeneye

Humans don't have golden eyes primarily because the pigments that color our eyes don't offer that variation. The golden eyes seen in other species occur because of the presence of a range of other pigments. In humans, there is only one exception to the melanin-fixed colorations: People with albinism (a lack of natural skin, hair, and eye pigment) may appear to have red eyes, but this is because of an absence rather than a presence of pigment, leading the red blood vessels at the back of the eye to become visible.

Do surgeons work best in a silent operating theater?

Is silence really the best aid to concentration? The majority of surgeons appear to think not. In the *British Medical Journal's* Christmas issue for 2014, a group of surgeons revealed that music is probably played more than 60 percent of operating time. But who gets to pick—and does it really help medical staff to keep cool heads and steady hands?

From Opera to Grime

Music has been around in the operating theater for a long time—the first phonograph was brought into surgery in 1914—supposedly to assuage the anxiety of the patient. More recently, it's become more

accepted than not that something will be playing over the often high volume of machines beeping or whirring, and the constant barrage of request, consultation, and commentary that's the standard background to most operations. Who gets to choose the playlist? Usually the most senior person in the room—and it's been found that having their favorite music playing actually helps surgeons to speed up on some procedures.

A Question of Taste

The last word should probably go to a surgeon—in an article in *The Guardian* in 2015, a number of surgical professionals were asked what they preferred to operate to, and their responses revealed a wide range of musical tastes, from opera to rock and rap. One orthopedic surgeon at King's College Hospital in London said, "If I play cool music it puts me in a better mood and I perform better. But music in theaters is a complex social issue . . . Last Christmas the anesthetist played a Christmas playlist for eight hours. Awful." And you might not have wanted to be the patient, either.

CAUSE AND EFFECT

Actions have consequences, and the human body is no exception to the rule. Discover whether you've registered all the connections by taking this quiz.

Questions

1. Humans produce two different kinds of tears. True or false?

2. Does having the words to describe something help you to remember it better?

3. What's the lifespan of the cells in your gut?

4. What new name for the evolutionary survivors previously known as "living fossils" was coined in 2014?

5. Learning to play the accordion is just one of the unusual solutions proposed for snoring—true or false?

6. Why are most Caucasian babies born with blue eyes?

7. Why do overweight people have a greater tendency to snore?

8. Are tears made of just water and salt?

9. Gargalesis, somatonesis, knismesis. One out of these three terms is invented—which one, and what are the other two?

10. Why is music often played in the operating theater?

Turn to page 216 for the answers.

WHICH INFECTIOUS DISEASE HAS KILLED THE MOST PEOPLE IN HISTORY?

HOW MUCH SLEEP DO YOU REALLY NEED (AND HOW MUCH IS TOO MUCH)?

COULD DRINKING URINE REALLY BE GOOD FOR YOUR HEALTH?

WHEN WAS THE FIRST OPERATION?

IN SICKNESS AND IN HEALTH

COULD YOU DIE OF PAIN?

Could drinking urine really be good for your health?

Every so often it crops up online—usually in the "ten trends you won't believe" clickbait column—Madonna, for example, has famously been reported to urinate on her feet to cure attacks of athlete's foot. But even under its technical name of urotherapy, it doesn't sound terribly appealing.

Where Did it Originate?

Using urine medicinally, both taken internally and applied topically, is genuinely an ancient practice (it's quoted in some of the earliest Sanskrit health texts), and has a place in the Ayurvedic and traditional Chinese medicine traditions. Urine was used as an antiseptic, and has been prescribed for a range of ailments, from kidney infections to cancer. Given its very long pedigree as a traditional cure-all, then, it's surprising that there

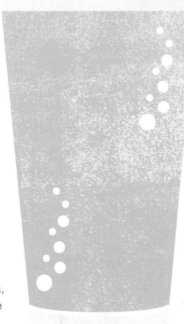

haven't been a greater number of significant research studies into its pros and cons as a form of medicine.

What's In It?

Here's what we know: Urine is 95 percent water, and 5 percent other substances—urea, minerals, enzymes, hormones, and salt. Although it's widely thought to be sterile on its exit from your body, this isn't necessarily the case, as your urethra, like your mouth and your gut, has its own colonizing bacteria.

No Harm, No Help

Western medical professionals have, on the whole, failed to give urotherapy a resounding vote of confidence. Many feel that a regular dose may not actually do you any harm, but that it isn't likely to help you much, either. Minor claims for the benefits of urine, as a tooth-whitener or a cure for skin problems, for example, are often acknowledged, although not warmly, by conventional practitioners in the West. Most, though, pour scorn on urotherapy adherents' more ambitious assertions—that drinking your own urine can help to cure cancer, for example—since they are clinically unproven. (If you're interested, the cancer cure idea works like this: Tumors release antigens into the body, and they pass into the urine. If you collect and drink your urine, the antigens go back into your body, and will encourage your immune system to produce antibodies that, in turn, will attack the cancer cells. Doctors have described the idea as "interesting," but it remains medically unproven.)

TAKING THE STING OUT

Maybe you won't opt to drink urine, but what about applying it to jellyfish stings? It's proven pain relief, right? Unfortunately not. When you're stung by a jellyfish, the stinging cells that line its tentacles leave nematocysts containing poison on your skin—even stings from jellyfish that aren't life-threatening are fantastically painful and irritating. The nematocysts are sensitive to chemical changes in their vicinity, so washing the area with anything other than seawater is likely to be counterproductive: It may provoke them to release any remaining poison into your skin. Urine and tap water have a different chemical composition to the seawater that's the poison cells' natural environment, which means that rinsing with either is more likely to do harm than good. If you've been stung by a jellyfish, rinse the area thoroughly with seawater (and follow up by checking with a doctor).

Could you die of pain?

Pain is useful for letting you know that all is not well with your body. What, though, if you were suffering from a constant and very high degree of pain and no one could do anything about it? Would the pain itself ultimately kill you?

A Built-in Avoidance System

In the developed world in the twenty-first century, we're generally not used to pain: There are enough doctors and painkillers to ensure that prolonged agony doesn't happen.

When pain becomes literally unbearable, though, your body offers an escape: The central nervous system, if it's overwhelmed with pain signals from your nerves, will cut out, and you will lose consciousness. In early operations, before the discovery of anesthetics, patients would often pass out from pain on the operating table.

Dying of Shock

Although pure pain can't kill you directly, it can cause your death through circulatory shock. This means that your body is placed in such trauma through pain that it can't ensure that enough blood or oxygen is circulated around your cells, which are quickly and permanently damaged. Your heart rate will rise, your breath will be quick and shallow, and you will sweat. Subjects often lose consciousness and die, sometimes of a heart attack. But the real cause of death is shock.

Painful Reckoning

It seems extraordinary, given how large it looms in our consciousness, that there's no objective way to measure the degree of pain you are suffering. James D. Hardy, a scientist at Cornell University, certainly thought so. In 1940, working with his colleagues Helen Goodell and Harold C. Wolff, he devised the Dolorimeter (named for the dol, a unit of pain in turn named after *dolor*, the Latin word for pain or sorrow).

The Dolorimeter measured pain from 0.5 to 10.5 dols, which Hardy and his partners intended to become a universal scale on which all pain could be assessed. The trouble was that it proved impossible to agree on the degree of pain that subjects were feeling when they were tested. The more the machine was used, the more it proved that there's no consensus

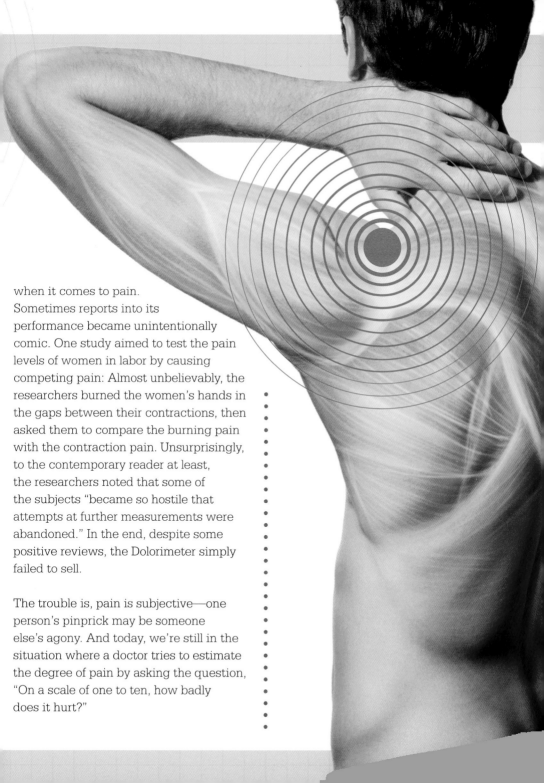

when it comes to pain. Sometimes reports into its performance became unintentionally comic. One study aimed to test the pain levels of women in labor by causing competing pain: Almost unbelievably, the researchers burned the women's hands in the gaps between their contractions, then asked them to compare the burning pain with the contraction pain. Unsurprisingly, to the contemporary reader at least, the researchers noted that some of the subjects "became so hostile that attempts at further measurements were abandoned." In the end, despite some positive reviews, the Dolorimeter simply failed to sell.

The trouble is, pain is subjective—one person's pinprick may be someone else's agony. And today, we're still in the situation where a doctor tries to estimate the degree of pain by asking the question, "On a scale of one to ten, how badly does it hurt?"

Which infectious disease has killed the most people in history?

The history of infectious diseases offers a grim roll call of mortality, and inevitably your mind goes to the last horrific outbreaks you heard of: Ebola virus, for example. But actually, Ebola ranks very low in what gallows humor might call the Grim Reaper awards.

How Much Do We Know?

Obviously it's impossible to say with absolute certainty which disease has been the deadliest across the whole of history, because we simply don't have the necessary records. Historical statisticians enjoy informed speculation, though, and their labors have turned up some interesting facts.

Take the Black Death, for instance—the wave of bubonic plague that laid waste to large areas of Europe, Africa, and Asia for seven long years in the mid-fourteenth century. Some historical statisticians reckon it caused a total of 200 million deaths, while others go down as low as a still-shocking 75 million. In Europe alone, it's estimated that around 50 million people, more than 60 percent of the contemporary population, died.

Tuberculosis vs. Malaria

Other major killers through history include influenza, cholera, the now-eradicated smallpox (which claimed its last victim in a laboratory accident in 1978), and malaria. But which killed the most? Over the last two centuries, a period for which it's easier to make accurate estimates than the more remote past, tuberculosis has caused an estimated 1 billion fatalities.

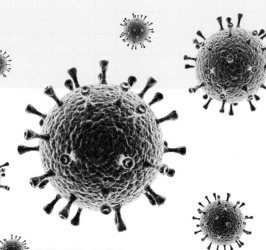

But malaria is still likely to be considerably out in front—for a while it was believed that it had killed half the people who had ever lived on Earth (with an estimate of 50 billion deaths). We don't have accurate enough information to tell us exactly how large the world's population was at different points in history to back up that claim—but experts believe the estimate could be close.

New Kids on the Block

The older diseases—and mosquitoes containing malarial parasites have been found dating back around 30 million years—have spent a very long time indeed claiming victims. At the other end of the scale, HIV and Ebola are both young diseases; neither has yet reached its fiftieth birthday. HIV is estimated to have killed around 36 million people since its emergence in the Democratic Republic of Congo in 1976. Ebola, which was "discovered" in the same year, has, despite its horrifying reputation, killed a comparatively tiny number of people overall; up to August 2018, fewer than 15,000.

OUTBREAK OR EPIDEMIC?

An "outbreak" indicates either an incidence of a greater-than-expected number of cases of an infectious disease in an area where it is already known, or one or more cases somewhere where it's never been seen before. An outbreak becomes an epidemic when the disease spreads to several areas from a common source, in much greater numbers than would usually be expected. For example, influenza may affect some people in a particular area every year, but when the numbers rise sharply, and the illness occurs over a larger-than-usual area, with several centers, it's become an epidemic. A pandemic is an epidemic on a global scale: It happens when an epidemic spreads across first countries and ultimately across continents.

How could antibiotics stop working?

Today it's hard to imagine life before antibiotics, even though the widespread use of penicillin goes back only to the mid-1940s. Their discovery transformed medicine, so it's frightening to be told, around 75 years after they were first introduced, that they're ceasing to be effective.

Natural Resistance

Antibiotics act against bacteria either by killing them outright or by preventing them from reproducing. And although it may seem to have developed very quickly since humans discovered how to make use of them, bacterial resistance to antibiotics is a natural occurrence. We're partly to blame ourselves for having used vast quantities of antibiotics in every imaginable situation: We've given the bacteria plenty of opportunity to practice.

Know Thy Enemy

Resistance happens because those bacteria least susceptible to the effects of antibiotics are the ones that survive and reproduce. It's survival of the fittest at a bacterial level. Over time, as only the toughest bacteria survive, they develop strains that respond less and less to antibiotics, until eventually they are fully resistant, and the antibiotics won't work. And some have developed multiple resistances to a range of antibiotics, creating "superbugs" that are very hard to treat. Some bacteria can actually destroy the antibiotic— they produce an enzyme called beta-lactamase, which breaks down penicillin, rendering it useless.

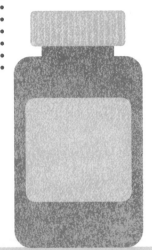

Looking Into the Future

Some authorities believe that we have only a very limited time before the majority of antibiotics become useless against the majority of bacteria (some have estimated just a decade or so). This would take medicine back to the point at which ordinary infections became killers, and it would be difficult, if not impossible, to perform operations safely. Apart from stringently limiting prescription of those antibiotics that bacteria find it hardest to fight against, to maintain their effectiveness for as long as possible, what can research do to protect us?

Researchers are looking at other ways in which our bodies might help us to protect ourselves. The immune system is the focus of a lot of work, in the hope that there might be a way to teach it to protect us better against bacteria in the first place, making antibiotics redundant.

COULD THE PROCESS GO INTO REVERSE?

It's possible that the resistance to antibiotics could gradually disappear, but only if bacteria didn't come into contact with them over a very long period. If the threat that antibiotics pose to bacteria were to vanish altogether, scientists believe that bacterial defense tactics would gradually also erode, but it could take a very long time without any antibiotic use at all—at least decades, and some say even centuries—before they would become effective again to any useful degree.

When was the first operation?

The first operations for which there is archeological evidence were trepanations, and they date back as far as late Paleolithic times—which means some may have been carried out around 12 millennia ago, and across a whole range of cultures, in Africa, the Americas, Asia, and Europe.

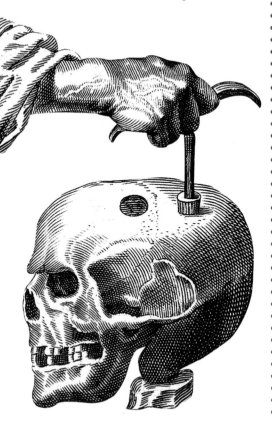

Holes in the Head

When you consider that a trepanation (or, as today's surgeons would call it, a craniotomy) involves literally drilling, or, worse, scraping a hole in your head, and that the only available tools in those early days would have been flints, or the sharp edges of seashells, the idea of undergoing one sounds less than appealing. So how and why were they carried out?

The first person to take an interest in an ancient skull with signs of trepanning was an American diplomat and archeologist called Ephraim (E.G.) Squier, who, in the 1860s in the course of a visit to Peru, was shown an ancient Inca skull bored with an evidently deliberate hole. He brought it home with him and it became the subject of intense debate: Had the subject survived the operation (had it, in fact, been performed pre-mortem at all)? Squier noted that the hole in the skull showed signs of having partially healed over. He concluded that it had been done on a living patient, and, moreover, that the patient had survived it.

Repair or Ritual?

The question "Why?" remained, and the debate broadened as, throughout the nineteenth and early twentieth centuries, more drilled skulls, many very ancient, were found all over the world. At the same time, scholars were pointing to the written works of the surgeons of the Ancient World, notably Hippocrates and Galen, both of whom advocated trepanning—originally named for the "trepan," or drill which was used for the operation in Ancient Greece and Rome—for some brain conditions and injuries. Even using a drill rather than a flint or a shell, it sounds tricky. Early writers urged that a bowl of water be ready to plunge the drill into when it became too hot, and commented that it was hard work to bore all the way through the skull without inadvertently going into the brain itself. From these sources, it might have been deduced that all trepanning was carried out for medical reasons. But while some skulls showed evidence of wounds or trauma, others seemed to have been carried out on skulls that showed no sign of previous damage, raising the question: Did trepanning have ritual purposes, too?

Letting in the Light

Trepanning has always involved letting something in or out of your head. When it wasn't a straightforward procedure to relieve pressure on the brain, it seems often to have been a case of letting devils out or enlightenment in. Aficionados—perhaps surprisingly, there are some contemporary enthusiasts—believe that it restores the natural lightness and well-being that you feel in childhood, before the bones of the skull fuse and harden. In 1970, Amanda Feilding, a British artist and drug policy reformer, performed the operation on herself while filming it. Afterward, she said, she wrapped her head in a scarf, ate a steak to replace the blood she'd lost, and went out to a party. She claimed that the mental benefits were subtle, but positive. In interviews she stuck with official medical advice: Don't try this at home.

How do diseases die out?

Generally diseases have proved able to resist human efforts to eradicate them—influenza, malaria, and measles, after all, are all still with us. The exception is smallpox, which was a scourge all over the world, killing millions, until it was finally wiped out in the 1970s.

Early Beginnings

Smallpox is a very ancient disease. When unwrapped, the mummy of Ramses V, dating back to the twelfth century BCE, was found to be marked with the scars of smallpox blisters. Of those who contracted it, 30 percent died, and even those who survived were often left blind and almost always with a disfiguringly scarred and pocked face.

The human fightback began with a process called variolation, of which the intrepid world traveler Lady Mary Wortley Montagu was an early advocate. Introduced to Europe in the eighteenth century, variolation was kill-or-cure risky: Pus was taken from smallpox blisters and someone who hadn't had smallpox either breathed it in or had it placed into a cut on their arm. Most then had an attack of smallpox, but fewer died than if they had caught it naturally. In 1796, Edward Jenner developed vaccination, a way of guarding against a disease by deliberately infecting subjects with a similar but milder disease. He had noticed that milkmaids, who had invariably had cowpox, never contracted smallpox, so began deliberately "vaccinating" (injecting subjects with a vaccine made from cowpox lesions). Once there was a vaccine, humankind had an effective weapon against smallpox.

The Death of Smallpox

Long after it had been eradicated in the richer countries of the world, smallpox raged on in the poorer ones. What finished it in the end was a global program of vaccination, accompanied by rigorous isolation policies around any sufferers. The World Health Organization announced a program of eradication in 1959, followed by one of "intensified eradication" in 1967. Eventually

COULD SMALLPOX RETURN?

smallpox survived only in Africa, where a stringent vaccination campaign eventually saw it clinging on in just three countries: Somalia, Ethiopia, and Kenya. The last natural case of smallpox occurred in Ethiopia in 1976, although there were to be a few further cases due to accident. Its eradication was announced by the WHO in 1980, in a triumphantly worded resolution: "The world and all its peoples have won freedom from smallpox . . ." it said, "[demonstrating] how nations working together in a common cause may further human progress."

Smallpox may have left the world stage, but it's not yet extinct—samples survive in a couple of laboratories, one in the United States, in the Center for Disease Control in Atlanta, Georgia, and the other in the State Research Centre for Virology and Biotechnology at Koltsovo, Novosibirsk, in Russia. Originally, the idea was to keep samples for research purposes in the event of breakouts from unforeseen sources; today, they've long been the subject of international wrangling. There's the fear of biochemical warfare, of course: In the wrong hands, smallpox would be the ultimate chemical weapon. And stores of the vaccine no longer exist. In 1990, the World Health Organization allegedly destroyed 9.5 million of the last 10 million doses on the grounds that they were no longer needed.

How much sleep do you really need (and how much is too much)?

When the media fixate on health issues, you hear quite a lot about how essential food and water are to survival, and in what quantities, but rather less about another essential in life—sleep. Most studies into insomnia in humans have been quite limited in scope, as sleep deprivation is an acknowledged form of torture.

The Effects of Sleeplessness

If you suffer from insomnia, you'll know that running short of sleep for even a single night can leave you feeling irritable, dopey, and unfocused. Experiments in sleep deprivation prove that low or no sleep is extremely bad for you. As little as 24 hours of sleeplessness has negative physical effects: First your blood pressure goes up, then, after around 48 hours, your body begins to stop dealing with glucose, your immune system becomes less effective, and your temperature drops. No one has ever gone the full mile on sleeplessness experiments on humans, but laboratory rats have been killed by imposed sleeplessness, though researchers couldn't agree on the physical cause of death.

The 8-Hour Mantra

Eight hours' sleep a night is fairly universally agreed to be the ideal for most people, keeping them functional and rested. But if too little sleep is bad, so is too much. You may also have experienced the situation in which a long lie-in actually leaves you feeling

more lethargic than before. The reason for this is that you're disrupting your circadian rhythm, the internal 24-hour clock that sets your body's habitual routine—it runs a tight ship and prefers, sleep-wise, to keep things regular.

Harvard University ran an extensive and lengthy study on the health of a large group of nurses and found that both over- and under-sleepers performed more poorly on memory tests than their well-rested colleagues—to the extent that the good sleepers (who were consistently getting between 7 and 8 hours' sleep a night) were judged to be 2 years younger, in memory and brain function than those who got either fewer or more hours of rest.

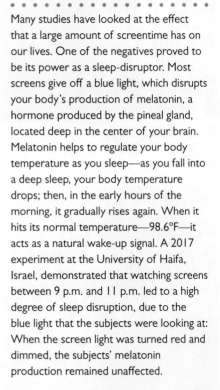

SEEING THE RED LIGHT

Many studies have looked at the effect that a large amount of screentime has on our lives. One of the negatives proved to be its power as a sleep-disruptor. Most screens give off a blue light, which disrupts your body's production of melatonin, a hormone produced by the pineal gland, located deep in the center of your brain. Melatonin helps to regulate your body temperature as you sleep—as you fall into a deep sleep, your body temperature drops; then, in the early hours of the morning, it gradually rises again. When it hits its normal temperature—98.6°F—it acts as a natural wake-up signal. A 2017 experiment at the University of Haifa, Israel, demonstrated that watching screens between 9 p.m. and 11 p.m. led to a high degree of sleep disruption, due to the blue light that the subjects were looking at: When the screen light was turned red and dimmed, the subjects' melatonin production remained unaffected.

Could traditional measures against plague have worked?

If you were unlucky enough to live in Europe when the Black Death arrived in the mid-fourteenth century (or during one of the many subsequent outbreaks of plague), professional medicine had little to offer by way of protection—it would be six centuries before antibiotics arrived to present a solution.

Four Useless "Cures"

Contemporary physicians, however, were still happy to develop a large number of "cures," many of them very expensive. Most sound unappealing; none were effective.

For the rich, a scoop of crushed emeralds might be prescribed. The stones were bashed into a gritty rubble using a pestle and mortar, then mixed with water and drunk down. Alternatively, you might try theriac. This well-known cure-all could contain up to a hundred ingredients, and there were many different recipes. Constants included opium, cinnamon, saffron, ginger, myrrh, and castor. Less conscientious quacks relied on treacle as a makeweight. Bloodletting and leeches were also popular—although they probably acted as the last straw for plague-weakened systems.

Quarantine

There was one measure that helped to slow the spread of the disease— although it did nothing for the unfortunates who had already caught it. It was the rigorous isolation of victims from the healthy. The practice of isolation wasn't invented in response to the Black Death (it was already in use for sufferers from certain diseases, such as leprosy), but it was considerably refined by it. It was newly set at 40 days for sufferers (the word quarantine developed from *quaranta*, the Italian word for forty); by the end of this time, anyone with plague would have either succumbed or, in very rare cases, recovered. Ports, through which plague had arrived, by way of the rats that passed freely on and off ships, were the first to introduce it— Dubrovnik led the way in 1377, and it was soon followed by others. Plague hospitals were set up, and carefully policed; most were at a little

FOWL CREULTY

Surely the most bizarre treatment of any suggested against plague was the Vicary method, named for Thomas Vicary, the English doctor who dreamed it up. (He lived in the sixteenth century, so missed the main event of the Black Death, but this didn't stop him creating the oddest-ever prescription for the many later outbreaks.) A live chicken had its bottom plucked of feathers, and was then applied, rear first, to a sufferer's plague sores, and strapped into place. Treatment was over when the chicken succumbed, though it might be moved to another site on the body if it proved unusually hardy. The Vicary method, though entirely useless, proved popular with plague sufferers in Tudor England.

distance from ports or towns, although near enough to enable sick people to be taken there easily. Often they were located on islands, to ensure the minimum possibility of contact. Ships' cargoes were given the same treatment: Goods and materials were taken to specified locations and left there in fresh air until it was considered that they had "fumigated" enough to have become safe.

Is organic food really better for you?

Yes. Or probably not. Although it depends on your definition of "better." Joking apart, the subject of organic food is such a hot potato that it would be very hard to arrive at a definitive answer. Some studies of the available information have concluded that the health benefits aren't really measurable.

Green Isn't Black and White

Indefinite conclusions don't mean that organic foods *aren't* good for you, just that it would be impossible, at the moment, to prove that you were "healthier" as a result of eating them, in either the short or the long term. The whole area of organics is a minefield. For every plus cited by organic producers, there's a counterargument. Since organic produce tends to sell at a higher price point than nonorganic produce, organic food enthusiasts are already likely to be economically comfortable and to pursue a healthy lifestyle, which makes it hard to single out the benefits that organics specifically might confer. Back in 2012, a huge U.S. study combining findings from many different smaller studies was made at Stanford University and couldn't point to any definite, measurable health benefits, although it did find that organic food contains lower levels of pesticide residues (contrary to what many people believe, some "natural" pesticides can be used on organic crops)—and it prompted a furious backlash.

But it's Better for the Environment, Right?

Not necessarily. A small area growing all-organic is likely to create a healthier environment than one growing nonorganic crops. But the current yield per acre of organic crops is much lower than that of nonorganic intensively farmed crops, so you'd have to use an awful lot more land to grow the same quantity of food. Some environmentalists believe that if you went all-organic, more land would need to be given over to farming; others don't agree. But "organic" in isolation isn't automatically an environmental win. Organic meat, too, is the most costly way of using land there is, in terms of "yield" per acre. It is true, though, that producers need to adhere to higher levels of animal welfare than is required by nonorganic farmers.

A Luxury Market

Around 90 percent of the market for organic food is, as you might expect, in North America and Europe—rich countries where people can afford to be fussy about what they eat. Organic food sales in the United States reached $49.4 billion in 2017, up from $29 billion just 7 years prior, and only $3.6 billion in 1997. Even allowing for the effects of inflation, it's clear organic has become big business over the last 20 years, with players who can spend a lot on marketing.

EATING LOCAL

There's one thing, though, that very few experts dispute: Whether or not you eat organic, it's good to eat as local as possible. Food that hasn't traveled far is fresher; it's easier to check where it comes from, and low "food miles" mean that it really is kinder to the environment. This makes it worth familiarizing yourself with your local market (and asking plenty of questions), whether the produce is organic or not.

Do burned foods cause cancer?

At some point during every summer barbecue, there'll be one nutritionally alert guest who will choose their moment to mention that they've heard that burned foods can give you cancer—causing everyone else to pause mid-mouthful. Is the rumor true, or is the rumormonger being alarmist?

Rogue Chemicals

That informative guest is referring to the fact that some burned or charred foods have been found to contain higher levels of a chemical called acrylamide. It's a natural chemical that is produced as a part of the Maillard reaction, which occurs when starchy foods are cooked at a high temperature until the proteins and sugars in them turn dark brown and produce complex, savory flavors. Acrylamide is only present in some types of food cooked in specific ways: Boiled foods don't contain it, and it's not found in raw foods, or dairy, or even meat or fish. It's particularly linked to carbohydrate-heavy foods such as potatoes or bread that have been grilled, baked, or fried.

Playing Safe

Acrylamide was only discovered to be present in food at all in the course of a study in Sweden in 2002—and it's known to be a carcinogen, a substance that can cause cancer in some animals because it causes damage to their DNA, a fact that was confirmed by laboratory studies on rats and mice. The jury is still out on whether or not this result could mean that it's also carcinogenic to humans in the

quantities in which they're likely to consume it, but most of the large health agencies are playing safe with their advice. The World Health Organization says that acrylamide is "probably" carcinogenic, while Cancer Research United Kingdom, despite having funded a study across a number of European countries that found no strong evidence for a link between the consumption of burned foods and cancer, still advise that carbohydrates be cooked to a golden color rather than the darker shades of brown. In the United Kingdom, the Food Standards Agency has even run a health campaign with the slogan "Go for Gold," advising against eating foods that have been cooked to the point of blackening (although it's arguable whether anyone really *prefers* burnt toast).

The Californian Coffee Scare

Never mind toast, what about coffee? Acrylamide has also been found to form naturally in the course of coffee beans' roasting, and in March 2018, a judge in California ruled that every cup sold in the state must carry a warning that it has the potential to cause cancer.

Critics have pointed out that the courts' requirement was nonsensical, because it would be impossible to prove. (It said that coffee shops wouldn't have to print the warning if they could prove that acrylamide would cause less than a single additional case of cancer in every 100,000 lifetime coffee drinkers.) So far, the presence of acrylamide in coffee doesn't seem to have shortened Californian coffee queues.

Will medicine ever be tailored to our DNA?

DNA is famously known as the "blueprint" that all living things have, but with an ever-increasing knowledge of the detail of how it works, what are the chances of doctors creating medicines that are a specific fit to our individual DNA profiles?

A New Science

To some extent, this is already possible. The comparatively new study of pharmacogenomics looks at ways to personalize medicine to maximize its effectiveness (and minimize drawbacks, such as bad reactions), using individual DNA. But although it is anticipated that within a decade or two, DNA profiling may become the norm in doctors' offices, personalized medicine is still at a relatively early stage.

Working with the Individual

In genetic terms, most of our drug treatments are still surprisingly crude. When you are told you have a specific illness, you're usually given drugs that would be given to most other people with the same condition. If the first drug doesn't work, your doctor will try a second one. If that doesn't work, a third. And so on, until you run out of options. Our genetic profiles will give doctors the knowledge, in advance, of which treatments are likely to work and which won't "gel" with our genes.

Cancer is at the forefront of personalized medicine research because so many cancers are the results of genetic mutations. In 2011, the *Wall Street Journal* reported the number of cancer tumors that were specifically the result of genetic mutations that could already be targeted by tailored treatments.

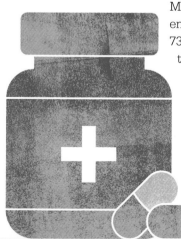

Melanomas got the most encouraging results with 73 percent of genetic tumors treatable, but even hard-to-treat lung and pancreatic cancers came in at 41 percent.

IN SICKNESS AND IN HEALTH

How much do you know about illness and health, past and present? This chapter goes far and wide, from the elimination of smallpox to whether or not organic food is really good for you—so take the quiz to find out how much you've absorbed.

Questions

1. Peeing on a jellyfish sting helps ease the pain—true or false?

2. What is pain for?

3. Ebola is a young disease. In which year was it identified and named?

4. Trepanation is believed to have been the first sort of operation ever performed. What is it?

5. In which country was the last natural case of smallpox found?

6. Why might banning screens from the bedroom be good for your sleep?

7. For which disease would a scoop of crushed emeralds have been prescribed as a cure?

8. Pharmacogenomics is a comparatively new science. What is it?

9. Which is better: eating organic food, or eating food produced locally?

10. What is acrylamide?

Turn to page 216 for the answers.

DOES YOUR WHOLE BODY DIE AT THE SAME TIME?

RIP

IS DEATH THE NORM FOR EVERY LIVING THING?

WOULD YOU HAVE ANY TIME TO THINK AFTER YOUR HEAD WAS CUT OFF?

ARE VENDING MACHINES DEADLIER THAN SHARKS?

DEATH AND AFTER

WHAT'S THE WEIGHT OF THE HUMAN SOUL?

What's the weight of the human soul?

To the contemporary thinker, it's an odd question: Even if people have souls, what reason is there to think that they can be weighed? But back in the early twentieth century, Dr. Duncan MacDougall of Haverhill, Massachusetts, felt that to weigh the soul would be to prove its existence.

The 21-gram Experiment

His subsequent, and very dubious, experiment led to a myth that persists to this day, and which even became the title of a movie, *21 Grams*, in 2003. To prove his point, Dr. MacDougall felt that all he would need to do was to weigh some human subjects immediately before and directly after death—he was certain that there would be a difference at the pre- and post-mortem weigh-in and that this difference would be the weight of the human soul, newly departed from the body.

21g

Weighing Things Up

His arrangements sound more like a story by Edgar Allan Poe than preparation for a scientific experiment. To carry out his study, he set up a hospital bed on sensitive scales. Somehow, he managed to recruit six dying patients, particularly favoring those who suffered from tuberculosis, as—he commented—they were lethargic, exhausted, and unlikely to move around much in their final hours, minimizing any accidental effect on the scales. His subsequent account of his subjects' dying moments shows a keen sense of drama. "Suddenly," he wrote, "coincident with death, the beam end [of the scales] dropped with an audible stroke, hitting against the lower limiting bar . . . The loss was ascertained to be three-fourths of an ounce."

He followed his dubious experiment up with another, this time using fifteen healthy dogs—how he induced them first to stay still, and then to die, remains unclear. He noted that there was no loss of weight after a canine death, and so it couldn't be considered that dogs had souls.

Heavy Personalities

MacDougall's article, published in 1907 and subsequently reviewed in the *New York Times*, nonetheless met with a skeptical response. Most scientists considered his sample size too small, and his data too spotty to be taken seriously. In particular, doubters asked why the loss of weight seemed to happen at varying times—did souls leave their bodies at different rates? Certainly, replied MacDougall—in "persons of sluggish temperament" it might take a little longer. Ultimately, the experiment was largely discredited. MacDougall was not put off—he moved on to an attempt to photograph the souls of the dying with the then-novel X-ray machine. Why 21 grams? Probably purely because it sounds more impressive than the much more nebulous "around three-quarters of an ounce"—which certainly wouldn't have made a good movie title.

A WEIGHING OF SOULS

MacDougall wasn't the first person to hold the conviction that souls could be weighed. The Ancient Egyptians believed that the jackal-headed god Anubis would test every human soul—sited in the heart—on scales against the feather of Ma'at, the goddess of truth. If the soul outweighed the feather, it wouldn't go on to the Field of Reeds, the Ancient Egyptian afterlife, but instead would be snapped up by the crocodile-faced Ammit, "gobbler of souls," who, in wall paintings and papyrus scrolls, is often shown waiting conveniently nearby, under the scales.

Would you have any time to think after your head was cut off?

You may have read about chickens standing upright and even running around for a time after their heads have been cut off. But what about people? If your head was cut off, would you have any degree of consciousness afterward?

Dead or Alive?

It's a hard question to answer definitively, as no one has yet returned from the dead to explain how decapitation felt. Historical accounts have often been lurid—and unbelievable. Charlotte Corday, the murderer of Jean-Paul Marat, a radical journalist during the French Revolution, is said to have blushed and turned to look at her executioner indignantly when he held up her head and smacked her face after she was guillotined. A more recent account is that of the French doctor Gabriel Beaurieux who, in 1905, witnessed the beheading of a man named Henri Languille, who had been convicted of armed robbery. When Beaurieux called Languille's name, he claimed that the head of the dead man had

opened its eyes and looked directly at him, not once, but twice. "I was dealing with undeniably living eyes that were looking at me," he recorded. Called for a third time, the head did not respond.

What Happens When You're Beheaded?

Could these stories be true? When your head is separated from your body, your circulation stops (after a few messy seconds, when blood will spurt from the carotid arteries in your neck). From the point at which your brain is no longer connected to your heart, fresh oxygenated blood will no longer reach it, but the blood that is already in your head may remain oxygenated for a few seconds, so theoretically you could remain conscious for that time.

As usual, the laboratory rat has been the fall guy when it comes to decapitation studies. (Sensitive readers might want to stop here.) In 2011, an experiment carried out at Radboud University in the Netherlands recorded the electrical activity in the brains of rats by wiring them up to an EEG machine to measure their brain waves—and then cutting their heads off. The EEG machine recorded activity at a level that implied consciousness, and probably conscious thought, for up to 4 seconds after death: Not very long, true, but enough to make uncomfortable reading, just the same.

LIKE A HEADLESS CHICKEN

Miracle Mike the Headless Chicken lived for a record 18 months after he'd been decapitated in a farmyard in Colorado in 1945. After his head was cut off, his body started to run around. His owner, Lloyd Olsen, placed the body in a box overnight, but when he checked in on Mike the next morning, he had a surprise: "The damn thing was still alive," said Troy Waters, Olsen's great grandson, retelling the story. Mike was tested and given a clean bill of health—to the extent that a headless chicken could be—by the University of Utah. Troy continued to feed Mike directly into his open esophagus with an eyedropper, and the headless chicken went on to earn a fortune for his owners when he was exhibited in touring sideshows. Mike eventually choked to death in a Phoenix motel room in 1947.

What's the best way to preserve a human body?

The best method of preservation depends on what use your body is to be put to when you're dead. Are you aiming to turn up to your religion's specific afterlife in good shape? Or do you want to explore the possibility of being brought back to life when humans have discovered how to defeat death?

Well-Preserved

Mummification has been practiced for millennia, most famously by the Ancient Egyptians. If you want to go to the afterlife with a well-preserved exterior it's a good choice, but for top levels of preservation your interior organs will need to be removed. After you've been eviscerated, things improve: Your body will be rinsed with wine, adorned with spices, and left in a thick layer of salts to dry out. For organ retention, a natural mummification in a peat bog may be the best option. The remains of Tollund Man, found in Denmark in 1950, were so well preserved that the peatcutters who found him thought at first that his body—which had been interred for around 2,000 years—was a

recent burial. Unlike the desiccated mummies of Ancient Egypt, peat bodies look more recognizable, their soft tissues as well as their skin and bones preserved in an extremely acid environment that contains hardly any oxygen.

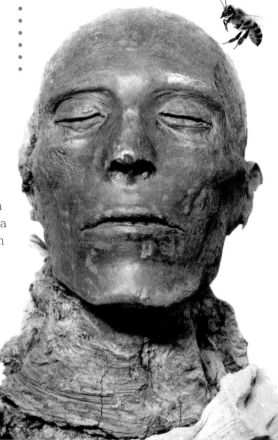

SWEET RELEASE

No preservation technique is odder than mellification—the practice of preserving bodies in honey. The technique was recorded in the *Bencao Gangmu (Compendium of Materia Medica)*, an extensive Chinese collection of medicines authored by the sixteenth-century apothecary Li Shizhen. Among the cures listed is one for broken bones: The sufferer should eat small pieces of mellified bodies—a sort of cannibalistic marrons glacés. As recorded, mellification was a spectacularly lengthy process that began while the subjects, usually elderly holy men, were still alive, by feeding them exclusively with honey. When they died (and diabetes must have been on the cards, given the sugar content of their diet), their bodies were immersed in huge jars full of even more honey and left there for a century. Only then were they removed and broken into small— and very pricey—pieces.

Embalming

Vladimir Lenin is probably the poster boy for this—well over 90 years after his death, the Russian leader is looking good (in fact, his remains have been constantly improved with an annual fresh embalming and cosmetic overhaul). Preparations for the public display of his body involve lengthy baths of, among other ingredients, formaldehyde, glycerol, and hydrogen peroxide. Some pieces have quietly been replaced: Artificial hair and eyelashes have been added, and when the body is displayed in public, his clothes go over a thin rubber "coat" to keep the chemicals close to his skin.

Frozen

Not the Disney hit, but the science of cryogenics: freezing dead bodies in the hope that they can be brought back to life in the future. As soon as you die, you'll be immersed in an ice bath to get your temperature down as rapidly as possible. Next, your blood will be drained and replaced with antifreeze before you're packed in ice and cooled further over the next fortnight until you reach the magic temperature of −320°F, after which you'll be stored in liquid nitrogen until scientific advances allow you to be either brought back to life or cloned.

Is death the norm for every living thing?

Threescore years and ten may be the biblical span allotted to humans, and many other familiar animals and plants have quite finite lives, but if you cast your net wider, you can find some kinds of life that don't seem to play by what you thought were the rules.

Tough As a Tardigrade

Tardigrades, tiny animals that live in a variety of mostly damp habitats, have a cartoon-like name and look like Pokemon personalities if you put them under the microscope—which you will have to do if you want a detailed look, as few of the thousand or so different species of tardigrade grow longer than a millimeter. But their cute appearance (solid bodies, squashed-up faces, and eight legs with clawed feet, all of which have led to their colloquial names —waterbears or moss piglets) is misleading: Tardigrades are one of the toughest life-forms in the world. Although their favored environment is a damp one—many species live in sediment at the bottom of rivers or lakes—tests have found that they seem to be very nearly indestructible in a wide variety of habitats and situations (they've even survived trips in space). One of their special talents is to self-dehydrate, bringing metabolic processes to a sharp halt to achieve a near-death state called cryptobiosis, which enables them to survive apparently indefinitely until conditions become more favorable.

Achieving Immortality

Other non-dying oddities include *Turritopsis dohrnii*, the immortal jellyfish, which can grow younger as well as older. For most of us, reproduction is a stop on the path to redundancy and, ultimately, death, but the immortal jellyfish lives up to its name by reproducing, then reverting to a polyp, the immature state in jellyfish. It then grows up all over again, reproduces and repeats the cycle. Ad nauseam.

The Undead

Viruses may be a slightly unfair entry on the list of immortals, as there's long been a scientific debate about whether or not they're alive in the first place. Some consider them simply line-ups of chemicals—but they challenge that definition when they hijack cells and get active. The cells (called hosts when they've become unwitting homes to a virus), are subjected to a full-on takeover, in the course of which the virus will induce them to reproduce its own DNA or RNA to make more virus. Smart thinking for mere chemicals.

CELLS THAT DON'T QUIT

Survival as a multicellular organism is one thing, but what about life at the cellular level? A sleeper hit in 2010, the book *The Immortal Life of Henrietta Lacks* documented the astonishing legacy of Lacks, who died of cervical cancer in 1951. Unknown to her—subsequently making for a difficult ethical position for her doctors—cells from her cancerous tumor had been taken while she was undergoing radiation treatment at Johns Hopkins Hospital and were subsequently used in lab research. One researcher, George Otto Gey, noticed how tough the Lacks cells seemed to be, outlasting all other cell samples, so he cultivated a line of cells from the originals, and named it the HeLa strain after Lacks. Nearly 70 years after her death, the HeLa strain is set to be truly immortal: still going strong, and used in laboratories all over the world.

❝ Tardigrades are one of the toughest life-forms in the world . . . One of their special talents is to self-dehydrate, bringing metabolic processes to a sharp halt. ❞

Do human bodies make good fertilizer?

Do you ever think about what happens to your body after you die? Both traditional burial and cremation have strong disadvantages when it comes to their green credentials and can remain toxic for the environment long after your death.

Burned or Buried?

If you opt for a traditional burial, you'll be contributing to the industrial quantities of metals and wood that annually go into the ground in the form of coffins, plus all the attendant waste, not least the millions of gallons of preserving fluid (usually a mixture of formaldehyde, methanol, and other solvents) that is used to stop bodies waiting for burial from decomposing, and is toxic for most of the natural life it will encounter in the soil when the coffin surrounding the body rots away. What about cremation—is that a greener solution? Arguably not: Your body will be burned at a very high temperature, which will release an estimated 540 pounds of carbon dioxide into the atmosphere, and rob your ashes of any potential nutritional value as fertilizer—subsequent scattering will be symbolic only.

Eco-Death

If you don't put your body through any of this, what's your nutritional worth to Planet Earth? A green burial will do away with embalming fluids (the vast majority of woodland or green burial sites specifically prohibit their use) and you'll be buried in an eco-coffin, made from materials such as card, bamboo, banana leaves, wicker, or willow, which will rot away far more speedily than a traditional wooden coffin. Most green sites offer the opportunity for the bereaved family to plant trees or flowers rather than using headstones; the ultimate aim being to offer a peaceful, sylvan retreat to visit and reflect. There are even eco-burial pods, like eggs, in development, that will be buried directly under a young tree's roots, with the idea that roots and pod will merge and that your body will, very directly, act as fertilizer for the tree.

Earth to Earth

There's a movement, though, that believes human bodies are full of good things, and could be composted much more directly than even the greenest burial currently practiced. For example, Recompose, a small pioneering organization in Washington State, is looking at literally turning human bodies into soil, by using traditional speed-composting methods within reusable pods. It claims that bodies can be returned to soil within 30 days, without using any substances that harm the environment—and it's currently planning a pilot program. Human compost may be the logical future of a green after-death service.

WASHED UP: THE GREEN CYCLE

If you don't want your body to be either buried or burned, there's a modern alternative, called alkaline hydrolysis, which will dissolve it instead. It's quite new, and has currently been legalized in parts of Canada and a number of states in the United States. It works rather like a washing machine: Your body is placed in a large, pressurized tank which is then filled with a highly alkaline mix of potassium hydroxide and water, heated to 572°F, and put through something like a lengthy wash cycle, which takes up to 4 hours. After the final rinse, only cleaned, separated bones remain: The solution has dissolved your flesh and connective tissue entirely.

When you hear about DNA being "degraded," what does it mean?

QUESTION 94 QUESTION

It was probably the success of *Jurassic Park* back in 1993 that pushed a question that up to this point had preoccupied only serious scientists into the mainstream: Could you really make a new dinosaur out of preserved dino DNA?

Your DNA is Broken

The idea of dinosaur blood taken from insects preserved in amber sounds extravagant, but it was actually based on a series of experiments and discoveries made in the same decade as the movie. Although some scientists claimed that they had managed to extract DNA that was 120 million years old, gradually these claims were debunked. Although the DNA molecule is huge and, in chemical terms at least, simple, it relies on many links, and it's these that break as the molecule ages. When enough links have broken, the DNA is technically useless—Helen Pilcher, author of a book about the "new science" of de-extinction, memorably wrote that it would be like "trying to construct the 5,195-piece Lego Star Wars Millennium Falcon from just a few bricks and the picture on the box." As soon as something living dies, its DNA is susceptible to damage. It's affected by enzymes from other organisms it comes into contact with, and oxygen, water, and sunlight can all damage and break the ladder-rung strands that are key to its structure.

Old, but Still Useful

So how long can scientists expect DNA to live? In 2012, an Australian study into DNA taken from the bones of a huge extinct bird, the moa, showed that it had a "half-life" of 521 years (meaning that by the end of that time frame, half of its links would still be complete). In the (relatively) complex math of science, this meant that half of the remaining half would take a further 521 years to degrade, half of that remaining half another 521 years, and so on—resulting in a calculation that determined that it would take 6.8 million years for DNA to be completely destroyed. If this calculation proves correct, it means that it's not an impossibility for feasible DNA to be found in samples around, say, a million years old.

THE DIFFICULTIES OF DNA

Originally, the unique "fingerprint" of an individual's DNA was thought to be the magic bullet that would make criminal convictions foolproof. It's only over time—the first crime to be solved by DNA profiling (the murder of Dawn Ashworth, a British teenager) was in 1986—that things have turned more complicated. As tests for DNA have evolved to be more and more sophisticated, it's become much harder for the presence of small quantities to unequivocally prove something—for example, it's been found that it's possible to transfer DNA from one garment to another in the course of a washing-machine wash, a process that an amateur might assume would destroy the DNA altogether. And now that minute amounts of DNA can be reliably identified, it's also more credible that they could have been deposited innocently: handed over on spare change, for example. Whatever its future, it seems certain that DNA evidence will become more rather than less complex to handle in the future.

QUESTION 95 QUESTION

Does your whole body die at the same time?

It may be comforting to think of death as instant, but it actually takes a little time to change status from living to, well, not. After your heart stops beating, oxygenated blood stops circulating and the cells most dependent on it die first.

Inside Out

The cells of your internal organs go fast (that's why it's crucial to harvest kidneys or a liver within at most 30 minutes of death), while your skin cells, at the other extreme, can live for much longer. If you were a skin donor, rather than a kidney donor, doctors would have up to 12 hours to take the donation.

The Light at the End of the Tunnel

What about the bright light so many people report moving toward at the moment of death (that's the ones who "came back")? There may be a scientific explanation. Your brain is one of the last parts of you to shut down, and when it is starved of oxygen, one side effect before you lose consciousness is tunnel vision. Combined with a sudden closedown as you black out, the effect might be very like moving down a tunnel toward the light.

WHY DOES DEATH SMELL SO BAD?

The horrible scent of putrefaction derives from a cocktail of chemicals, most of them by-products of bacteria that move in to colonize a cadaver. Two of the smelliest culprits are the evocatively named putrescine and cadaverine, both naturally produced by the breakdown of amino acids in the body after death. One lab worker suggested that a quick squirt of the two combined and you'd smell "dead" enough to ensure your survival in a zombie apocalypse.

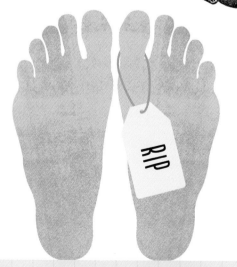

RIP

Are vending machines deadlier than sharks?

It's a comparison that was all over the internet a few years back, when it was mentioned in the commentary for visitors to the shark-and-ray tank at Boston's New England Aquarium. And statistical analysts confirmed that it is actually true.

Everyday Dangers

Vending machines kill an average of two Americans every year, while sharks are only responsible for a single death every 2 years (even the global number of deaths resulting from shark attacks is a modest-sounding annual average of between four and six victims).

Vending machines won't go on the attack, it's true, but they weigh an average of around 900 pounds and they can tip. In the cases of the few deaths that do occur, that's what usually happens—they fail to dispense a paid-for snack, are pushed and shoved in an attempt to dislodge it, and fall forwards.

To be killed by a shark, though, you have to be not only at the beach, but actually swimming in the sea—so you're not comparing like with like.

Cause of Death: Unexpected

The Centers for Disease Control and Prevention in Atlanta, Georgia, keep records of the causes of deaths in America. As well as vending machine fatalities, the records cover the statistics for other causes of death that you might find surprising. Between 1999 and 2014, 1,413 people died falling out of trees (that's an average of more than ninety-four a year), while power lawnmowers were the demise of 951 people (more than sixty-three deaths a year). So be careful of your lawnmower, your tree-climbing habits, and even your vending machine visits: All three are considerably more dangerous than sharks.

Does your hair keep growing after you die?

A number of ghost stories depend upon it, not least M. R. James's splendidly creepy *The Diary of Mr. Poynter*, which concludes (spoiler alert) with a coffin, "quite full of hair . . ." How much truth is there, though, in the frequently repeated story that, after you die, your hair will keep on growing? And if it's true, how long will it grow for?

The Root of the Matter

The hair on your head (and your chin and cheeks, if you have a beard and sideburns) is already dead—the only area of growth on each hair is at the base of the follicle. Where the hairs are rooted into your scalp, each one has a number of protein cells that enable growth. These cells depend on energy, derived from your body burning glucose, to stay active. And the supply of glucose depends on oxygenated blood pumping around your body. When your heart stops, so does your circulation, so your hair would continue to grow for only the few moments between your heart ceasing to beat and your circulation stopping.

What About Nails?

Your fingernails are also often thrown into the story—it's often alleged that they'll carry on growing after you die, too. Not true; just as your hair does, your finger- and toenails require a supply of oxygenated blood to keep growing. However, the desiccation of your skin after death causes the cuticles to begin to shrivel back from the nail bed— which can give the mistaken expression that your nails have continued to grow after the rest of you has ground to a permanent halt.

DEATH AND AFTER

You may no longer be aware of them, but plenty of things go on happening to your body after death. Check in on your knowledge of post-mortem facts with this handy quiz.

Questions

1. Which Ancient Egyptian god was responsible for weighing the human soul?

2. If your body was subjected to mellification after death, what would have happened to it?

3. How much carbon dioxide is released in the course of the average cremation?

4. Fetidine, putrescine, and cadaverine: two are real, and one is made up. Which is the invented one, and what are the other two?

5. Why is Henrietta Lacks remembered in research labs all over the world?

6. What are two common names for the tiny animals called tardigrades?

7. What was unusual about the chicken who came to be known as Miracle Mike?

8. How long is it estimated that DNA can last?

9. How soon after its owner's death does a donated liver need to be harvested?

10. Globally, ten people are killed by shark attacks every year. True or false?

Turn to page 217 for the answers.

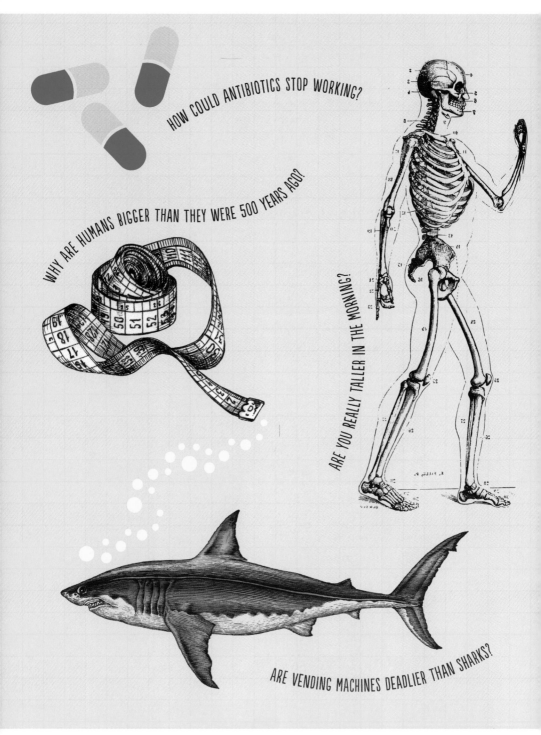

HOW COULD ANTIBIOTICS STOP WORKING?

WHY ARE HUMANS BIGGER THAN THEY WERE 500 YEARS AGO?

ARE YOU REALLY TALLER IN THE MORNING?

ARE VENDING MACHINES DEADLIER THAN SHARKS?

COULD YOU LIVE ON A VAMPIRE DIET?

HOW BIG IS A HUMAN EGG?

QUIZ ANSWERS

WHERE DO DEAD BODY CELLS GO?

WHY CAN'T YOU TICKLE YOURSELF?

QUIZ ANSWERS

Birth and Before Quiz Answers

1. Crown–rump length.

2. A baby has ninety-four extra bones.

3. Yes.

4. False. A human egg is around sixteen times the size of a sperm.

5. Because she has to birth her young through a very narrow tube.

6. False. Relaxin is the hormone that makes muscles and ligaments stretchier in pregnancy.

7. Apollo.

8. To ensure that people give her priority when standing in line, and a seat on the subway.

9. The first bowel movement of a newborn baby.

10. September—though not by much.

Fastest, Longest, Greatest, Strongest Quiz Answers

1. False. A bullet reaches around 1,700 mph, but the fastest sneeze speed ever recorded was only 102 mph.

2. Yes.

3. Your brain.

4. The liver.

5. Shed human skin cells, which make up a lot of house dust, contain an oil called squalene that absorbs the pollutant ozone.

6. An additional sense organ located just inside the nose of many mammals, notably dogs.

7. Taller.

8. The nervous system.

9. Arachnids.

10. Because it works so hard.

Historical Bodies Quiz Answers

1. b) Short.
2. The Denisovans.
3. True. Wolves prefer familiar foods.
4. They were (allegedly) turned into buttons.
5. Chimpanzees and rats.
6. King Charles VI.
7. Uttar Pradesh.
8. False. He published a treatise against the evils of tobacco in 1604.
9. Absolutely not.
10. There's no fossil evidence that people ever lived in water.

Fashionable Bodies Quiz Answers

1. Because fashionable and high-born Japanese women had dyed their teeth black for centuries.
2. The coloring of the white of the eye with ink.
3. They helped prevent riders' feet from slipping out of the stirrups.
4. Golden lotus.
5. Dress fabric and wallpaper (toothpaste sometimes did, too).
6. True. He believed that beards were too easy to grab in hand-to-hand combat.
7. Pierre-Joseph Desault.
8. They help keep foreign matter out of your eye, and break up the flow of air around your eyeball.
9. False. They developed metal razors, usually made from copper, to shave with.
10. A slanted Cuban heel, up to 2 inches high.

Internal Affairs Quiz Answers

1. Hyponatremia.
2. Two-thirds of it.
3. Possible answers: carbon dioxide, hydrogen sulfide, methane.
4. A cucumber.
5. Apoptosis.
6. Between 2 and 4 pounds.
7. False. Your lungs have a higher water content than your blood.
8. b) Small protrusions.
9. The potential medical applications for the microorganisms that occur naturally in your gut.
10. b) Short time window.

Unexpected Events Quiz Answers

1. Marie Antoinette.
2. In the saliva of a vampire bat.
3. Charles Dickens and Nikolai Gogol.
4. Ireland.
5. Admiral Horatio Nelson.
6. It would boil because the vacuum of space pushes the boiling point of fluids down below the natural human body temperature.
7. Kifuka, Democratic Republic of Congo
8. b) Eustress.
9. False. It's the point at which your ulnar nerve passes through your elbow.
10. Synesthesia.

In Your Head Quiz Answers

1. b) In possession of a super-memory.

2. He created the cartoon character Wonder Woman.

3. Dopamine.

4. Congenital insensitivity to pain.

5. Your brain takes mental snapshots of the view as your eyelid goes down and the view as it goes up again, then bridges the brief gap between the two.

6. False. It's the technical term for a so-called "ice-cream headache," caused by eating very cold substances quickly.

7. Possibly. Some studies have shown that the odds for developing Alzheimer's reduce with every additional year of education beyond primary school.

8. Use your natural sense of direction.

9. True. Referred pain can mean that a liver problem is experienced as neck pain.

10. It means the sleeper is dreaming.

Cause and Effect Quiz Answers

1. False. There are three kinds of tears: basal tears, reflex tears, and emotional tears.
2. No.
3. 3 to 4 days.
4. Stabilomorphs.
5. False. But the controlled breathing required to play the didgeridoo has proved a useful way to help with problem snoring.
6. Because they haven't yet acquired their full complement of melanins, the pigments that will "fix" their eye color.
7. Fat deposits in their airways partially obstruct the air flowing in and out of their lungs, resulting in an uneven-bellows pattern—snoring—in their breathing.
8. No. They also contain oil, mucus, and lysozymes, a natural form of antibiotic.
9. Somatonesis is invented. The other two terms both refer to types of tickling sensation.
10. It helps surgeons to concentrate.

In Sickness and in Health Quiz Answers

1. False. Stings are best washed with plain seawater.
2. It is an alarm system to warn you that something is harming your body.
3. 1976.
4. The creation of a hole in the skull to relieve pressure on the brain.
5. Ethiopia.
6. Screens emit blue light, which disrupts the production of the sleep hormone melatonin.

7. The Black Death (bubonic plague).

8. The study of personalizing medicine using a patient's individual DNA, through DNA profiling.

9. The health benefits of organic food aren't proven, while locally produced food won't have the environmental burden of "food miles" and is also likely to be fresher.

10. Acrylamide is a chemical that forms naturally when starchy foods are grilled, baked, or fried at high temperatures.

Death and After Quiz Answers

1. The jackal-headed Anubis.

2. It would have been preserved in honey.

3. 540 pounds' worth.

4. Fetidine is invented. Putrescine and cadaverine are real: Intensely smelly chemicals produced by the breakdown of amino acids in a dead body.

5. Because of the extraordinary longevity of her cells.

6. Waterbears and moss piglets.

7. He lived for 18 months after his head was cut off.

8. 6.8 million years.

9. Within 30 minutes.

10. False. Annually, sharks kill between four and six people in the world.

INDEX

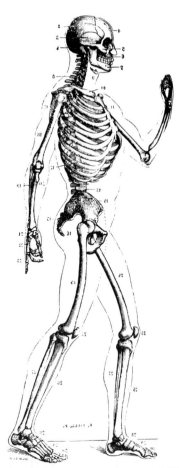

CREDITS